Community Energy and Sustainable Energy Transitions

Vanesa Castán Broto
Editor

Community Energy and Sustainable Energy Transitions

Experiences from Ethiopia, Malawi and Mozambique

palgrave
macmillan

Editor
Vanesa Castán Broto
Sheffield, UK

ISBN 978-3-031-57940-0 ISBN 978-3-031-57938-7 (eBook)
https://doi.org/10.1007/978-3-031-57938-7

This Palgrave Macmillan imprint is published by the registered company Springer Nature Switzerland AG
The registered company address is: Gewerbestrasse 11, 6330 Cham, Switzerland

Paper in this product is recyclable.

Acknowledgements

This book is the result of teamwork—including a great deal of mutual care, encouragement, and support—over five years and across four countries. The team was constituted with a grant from UK Research Innovation (UKRI) for the project *Community Energy and Sustainable Energy Transitions in Ethiopia, Malawi, and Mozambique* (CESET) (ES/T006358/1). Over the years, we have reflected our work in our website (https://cesetproject.com/), but we hope this book will be part of the legacy of the project. Working with the CESET team has been one of the most wonderful experiences any academic can wish for. Sometimes, we made each other believe that anything was possible. Other times, we supported each other through frustrations and disappointments—and through some very difficult times. This book is our collective present to each other: a memory of the time we have worked together and a statement of the kind of interdisciplinary, engaged work we hope to continue doing together.

The work presented in this book was made possible through a grant from the UK Global Challenges Research Fund (GCRF), managed by the Economic and Social Research Council (ESRC) of UKRI. The GCRF was an innovative research initiative to address the UN Sustainable Development Goals, which built research capacity across lower income countries. GCRF challenged modes of extractive research that were prevalent across academia ca. 2015 and sought to foreground the importance of equitable research partnerships, in-country demand, and building research capacity.

This funding programme transformed cultures and perceptions of the role of research in a postcolonial world, although there is still work to be done to address the numerous epistemic injustices that pervade contemporary academia. As a team, we would like to thank everyone in UKRI who made the GCRF possible, including the incredibly supportive Marzena Bien.

CESET started in April 2020 as the COVID-19 pandemic enfolded the world. The collaboration built over the writing of the proposal kept the project going, but many of us did not meet in person until we were able to hold the first project meeting in March 2022. Then, in November 2020, the Tigray War started. Some estimates calculate that up to 600,000 people died because of the war between 2020 and 2022. CESET partners at Addis Ababa University and Mekelle University were overwhelmed by the dynamics of war. The ESRC (thank you again, Marzena!) was incredibly supportive at this stage, as well as the support teams in both institutions.

Some CESET members have not been able to contribute directly to the book for either personal reasons or changing job demands, even though they have supported the project since its inception, including Yacob Mulugetta, Marcus Power, Joshua Kirshner, Domingos Augusto Maculule, Sofia Saguate, Martin Mganyasi, Sandy Robinson, Eve Dixon, Abdoumaliq Simone, and Collen Zalenguera. We all hope the work presented here reflects their perspectives too. Thank you to all the communities that have welcomed the project and engaged with us as a team, in particular, the people of Chamanculo who took us on a tour of the site and joined us on a big celebratory event in November 2022.

The project has been supported in the shadows by a range of professionals, including Anya Rayner, Erika Conchis, Dan Farley, and Sean Lovell. The wonderful Enora Robin played a key role in the early stages of the project before she went on to excavate the mysteries of alternative finance for solar energy. There is one person without whom the project would not run. This person is Vicky Simpson. If we are the players in the energy science game, Vicky is our coach. The number of 'thank yous' owed to Vicky for her diligence and capacity to anticipate challenges could fill another book.

We have had several companions in our CESET project, so many that it is impossible to be comprehensive. The GESIS group in Mozambique has been incredibly inspirational, especially Fatima Arthur, Clementina De Sousa, and Nirza Fumo, alongside our colleagues from the Mozambican government, Angelina Bombe and Olimpio Zavale. The experience in Mozambique has helped the CESET team to create strong linkages with other gender and energy experts, especially Joy Clancy, Saska Petrova, and Ralitsa Hiteva, who have all advanced CESET's thinking with their ideas and dynamic energy. The Transforming Energy Access Learning Partnership (https://tea-lp.org/) provided support to advance the education-related debates within CESET, especially Jiska De Groot, who provided advice to the CESET team a few times. Finally, our colleagues at the Modern Energy Cooking Services project (https://mecs.org.uk/) have provided unwavering support and inspiration. Ed Brown and Louise Medland, in particular, energised us and resolved a few bureaucratic complications.

There are so many other people to thank, not least all the wonderful people at each of the institutions participating in this book and the staff at the Urban Institute and the Faculty of Social Sciences at the University of Sheffield who have hosted the project. Catherine Barlow revised the book and made its index for submission to Palgrave. At Palgrave, the book was supported by Rachael Ballard and the immensely patient Agila Muruganandam.

None of us could do any of this work without the unsung support of our families, friends, and all of those who believe in our work. It is just too difficult to list all of you, but you all know who you are! Thank you!

This book is dedicated to all those who work to deliver universal access to clean energy in ways that work for people and in the manner in which people want it.

Contents

List of Contributors

Amare Assefa Adis Ababa University, Addis Ababa, Ethiopia

Idalina Baptista University of Oxford, Oxford, UK

Getachew Bekele Addis Ababa University, Addis Ababa, Ethiopia

Vanesa Castán Broto University of Sheffield, Sheffield, UK

Maxon L. Chitawo Mzuzu University, Mzuzu, Malawi

Carlos Alberto Cuvilas Faculty of Engineering, University Eduardo Mondlane, Maputo, Mozambique

Aran Eales University of Strathclyde, Glasgow, UK

Mulualem G. Gebreslassie Center of Energy, Ethiopian Institute of Technology-Mekelle, Mekelle University, Mekelle, Ethiopia, Materials and Engineering Research Institute, Sheffield Hallam University, Sheffield, UK

Chrispin Gogoda Mzuzu University, Mzuzu, Malawi

Dawit Habtu Adis Ababa University, Addis Ababa, Ethiopia

Christopher Hara Mzuzu University, Mzuzu, Malawi

Lorraine Howe Centre for Research on Governance and Development, Maputo, Mozambique

Arnold Kadziponye Mulanje Energy Trust, Mulanje, Malawi

Adugnaw Lake Adis Ababa University, Addis Ababa, Ethiopia

John Sailence Chipopoma Power, Livingstonia, Malawi

Carlos Shenga Centre for Research on Governance and Development, Maputo, Mozambique

Long Seng To STEER Centre, Geography and Environment, School of Social Sciences and Humanities, Loughborough University, Loughborough, United Kingdom

Hita Unnikrishnan Urban Institute, University of Sheffield, Sheffield, UK

Harshit Vallecha STEER Centre, Geography and Environment, School of Social Sciences and Humanities, Loughborough University, Loughborough, United Kingdom

Acronyms

ADLI	Agricultural Development Led Industrialization (Ethiopian development strategy)
ALER	Associação Lusófona de Energias Renováveis (Renewable energy association for Portuguese-speaking countries)
AMDA	African Minigrids Development Association
AMER	Associação Moçambicana de Energias Renováveis (Renewable energy association in Mozambique)
ARENE	Autoridade Reguladora de Energia (Energy regulation authority in Mozambique)
ARPU	Average Revenue Per User
ATM	Autoridade Tributária de Moçambique (fiscal authority) (Mozambique)
CAPEX	Capital Expenditure
CARD	Churches Action in Relief and Development
CARRI	Community and Regional Resilience Institute
CE	Community Energy
CEM	Community Energy Malawi
CEOSC	Community Energy Service Companies
CES	Community Energy systems
CESET	Community Energy and the Sustainable Energy Transition in Ethiopia, Malawi and Mozambique
CNELEC	Conselho Nacional de Electricidade (National Electricity Council) (Mozambique)
COP28	Conference of Parties, Dubai 2023
CRGE	Climate Resilient Green Economy (Ethiopia)
CRT	Cost-Reflective Tariff

DBE	Development Bank of Ethiopia
DFID	Department for International Development (UK)
DoE	Department of Energy Affairs
DRC	Democratic Republic of the Congo
EASE	Rural Energy Access Through Social Enterprise and Decentralisation Project
EDM	Electricidade de Moçambique (Mozambique's Electricity Utility)
EEA	Ethiopian Energy Authority
EELPA	Ethiopia Energy Light and Power Authority
EEP Africa	Energy and Environment Partnership Trust Fund
EEP	Ethiopian Electric Power
EEU	Ethiopian Electric Utility
EGENCO	Electricity Generation Company (Malawi)
EPA	Environmental Protection Agency (US)
ESCOM	Electricity Supply Cooperation of Malawi
ESMAP	Energy Sector Management Assistance Program
EU	European Union
FDRE	Federal Democratic Republic Government of Ethiopia
FiT	Feed-in Tariff
FUNAE	Fundo de Energia (Mozambique's Energy Fund)
GCRF	UK's Global Challenges Research Fund
GEF	Global Environmental Facility (UNDP)
GIZ	Agency for Technical Co-operation (Germany)
GoM	Government of Malawi
HCB	Cahora-Bassa Hydroelectric Dam (Mozambique)
HEP	Higher Education Proclamation (Ethiopia)
IEA	International Energy Agency
INNOQ	Instituto Nacional de Normalização e Qualidade (National Institute for Standardization and Quality) (Mozambique)
IoT	Institute of Technology
IPCC	Intergovernmental Panel on Climate Change
IPPs	Independent Power Producers
IRENA	International Renewable Energy Agency
kW	kilowatt
kWh	kilowatt-hours
MAREP	Malawi Rural Electrification Program
MDCL	Market Development Credit Line (Ethiopia)
MEF	Ministro da Economia e Finanças (Ministry of Economy and Finance) (Mozambique)
MEGA	Mulanje Energy Generation Agency
MERA	Malawi Energy Revenue Authority
MFI	MicroFinance Institution
MFIs	Microfinance Institutions

MIREME	Ministério dos Recursos Minerais e Energia (Mozambique's Ministry of Mineral Resources and Energy)
MMCT	Mulanje Mountain Conservation Trust
MoSHE	Ministry of Science and Higher Education (Ethiopia)
MoWIE	Ministry of Water, Irrigation and Energy (Ethiopia)
MSSD	Most Similar System Design
MTF	Multi-Tier Framework
MuREA	Mulanje Renewable Energy Agency
NEP	National Electrification Programme (Ethiopia)
NES	National Electrification Strategy (Ethiopia)
NGO	Non-Government Organisation
NREL	National Renewable Energy Laboratory (United States)
ODA	Overseas Development Assistance
OECD	Organisation for Economic Co-operation and Development
OPEX	Operational Expenditure
PAE	Pacote de Aceleração Económica, Mozambique
PAYG	Pay as You Go
PPA	Power Purchase Agreement (Mozambique)
PPP	Public–Private Partnership
PROBEC	Programme for Basic Energy Conservation (Malawi)
PSEs	Private Sector Enterprises
PUE	Productive Uses of Electricity
PV	Photovoltaics
RE	Renewable Energy
REF	Rural Electrification Fund (Ethiopia)
RENAMA	Renew'N'Able Malawi
SADAC	Southern Africa Development Community
SDGs	Sustainable Development Goals
SE4RC	Sustainable Energy for Rural Communities
SHS	Solar Home System
SNNPR	Southern Nations Nationalities People's Region (Ethiopia)
UCBP	University Capacity Building Program (Ethiopia)
UEAP	Universal Electrification Access Programme (Ethiopia)
UN	United Nations
UNDP	United Nations Development Programme
UNESCO	United Nations Educational, Scientific and Cultural Organization
UNFCCC	United Nations Framework Convention on Climate Change
UNIDO	United Nations Industrial Development Organization
USAID	United States Agency for International Development
USD	US Dollars
VAT	Value Added Tax
VEC	Village Electricity Committee (Malawi)
WASHTED	Water Sanitation, Health, and Appropriate Technology Development
WB	World Bank
WHO	World Health Organisation

List of Figures

List of Tables

CHAPTER 1

Introduction: Community Energy and Sustainable Energy Transitions

Vanesa Castán Broto

Abstract This chapter introduces the book on Community Energy and Sustainable Energy Transitions in Ethiopia, Malawi, and Mozambique. The book is the result of an interdisciplinary collaboration between an international team of scholars brought together thanks to research funds from the UK's Global Challenges Research Fund. The Introduction argues the need for a perspective from these three countries as a means to challenge existing assumptions about the development of community energy projects and their value for accelerating the transition to sustainable energy. The aim is to develop a postcolonial perspective on community energy that emerges from practical experiences on the ground and the challenges raised by those experiences. The chapter concludes with an overview of the book and a call to consider community energy as an inherently diverse phenomenon.

Keywords Community energy · Sustainable energy transitions · Social change · Energy infrastructure

V. Castán Broto (✉)
University of Sheffield, Sheffield, UK
e-mail: v.castanbroto@sheffield.ac.uk

V. Castán Broto (ed.), *Community Energy and Sustainable Energy Transitions*, https://doi.org/10.1007/978-3-031-57938-7_1

1.1 Introduction

In a recent essay on the need to recognise the 'slowness' of climate action in delivering results, writer Rebecca Solnit (2024) recalls Greta Thunberg's speech to British MPs, in which she embodied the frustration of many climate activists:

> Avoiding climate breakdown will require cathedral thinking. We must lay the foundation while we may not know exactly how to build the ceiling. (Thunberg, 2019)

The transition to sustainable energy may as well be a cathedral as it requires both a commitment to work over many decades and a firm belief in the transcendence of the effort. However, as Solnit explains, the cathedral of our common future will not be built only in grand plans but also in day-to-day actions that redefine whole cultures and societies because "a change of regime isn't a change of culture and consciousness" (Solnit, 2024: n.p.). Creating long, deep, lasting change requires a myriad of actions, from the mundane to the extraordinary.

For many scholars and activists participating in this book, building a cathedral is an apt metaphor for the task at hand. We have been brought together by a shared belief that putting communities at the centre of action is the most effective and fair means of delivering a transition to sustainable energy. At the same time, we realise that community energy is not yet fully recognised as a transition option outside very specific contexts, such as remote rural areas without grid coverage where no other workable alternative is available. The transition is our cathedral. Community energy helps to build part of its foundations. Committing to community energy requires a leap of faith and the capacity to imagine a very different energy future.

Such a leap of faith has brought together a group of interdisciplinary scholars to investigate the possibilities, advancement, and challenges for community energy. The project Community Energy and the Sustainable Energy Transition in Ethiopia, Malawi, and Mozambique (CESET) is a collaborative multi-institutional research partnership funded by the UK's Global Challenges Research Fund that explores the potential of community energy systems to accelerate inclusive, just, and clean energy transitions in Ethiopia, Malawi, and Mozambique. Putting the research focus on these three countries enables examining community energy contributions to the combined challenges of providing universal energy access, enabling the decarbonization of energy systems, and increasing communities' resilience in the face of increasingly tangible climate impacts.

A recent report from a panel of African experts on the Just Transition argues that "*new energy systems need to be powered by a large diversity of actors, not least common good oriented and non-profit maximising entities such as households, farmers, cooperatives, community associations, schools, universities, hospitals and other public service entities*" also embracing diverse-sized renewable projects and mechanisms to democratise energy systems (Sokona et al., 2023: p. 50). This will require rethinking infrastructure models of energy development away from the model of the grid but also piloting multiple institutional arrangements for governance and ownership, such as cooperatives and partnerships, alongside the participation of the government and the business sector.

The experiences of the CESET countries have much to contribute to this debate. Ethiopia has pioneered community energy experiences, but the community energy model is still seen as a secondary strategy for energy access. Malawi has embraced community energy as a workable alternative to deal with the country's enormous gap in access to electricity, but limited resources, the pressures of foreign exchange rates, and fragmented supply chains have stalled the development of projects. Mozambique adopted a new off-grid regulation in 2021 that could favour the entry of community energy, but the dominance of the public sector on infrastructure development and the emphasis on capital investments constrain the imaginations of what community energy can achieve in that country. Learning from these experiences, a key research question is: what is the potential for community energy to catalyse not just a technological but also a social, institutional, cultural, and economic transition to sustainable energy?

Community energy forces a revision of standard thinking about transition and whether transitions can be just. Zero-carbon climate-resilient futures must be addressed alongside social development programmes that address differential vulnerabilities that shape communities' sensitivity to climate impacts and their capacity to respond (Lee et al., 2023). One response has been to call for transitions that incorporate justice as a core component of the actions to change energy sectors. Unfortunately, debates on the just transition have been dominated by an aspiration to develop conceptual clarity (e.g., Heffron & Heffron, 2021; Wang & Lo,

2021) at the expense of delivering practical responses tackling the drivers of oppression of people and the world—the drivers that lead to climate breakdown in the first place.

While transitions are often thought of as transforming the status quo, incumbent institutions—those who hold power—often adapt to changing conditions, shifting narratives to claim new forms of legitimacy (Novalia et al., 2021). Sometimes, innovative technologies become the instruments for elites to build such legitimacy (Swyngedouw, 1997). Often, lofty discourses of justice overlook the complex trade-offs that shape the political economy of energy and any transition occurring within it (Newell & Mulvaney, 2013). For example, transitions have different impacts at a distance, including the reconfiguration of supply chains for energy projects, which may lead to the creation of new 'green sacrifice zones' to satisfy new needs of mining and resource exploitation within the renewable energy industry (Zografos, 2022). There is, thus, great pessimism about the extent to which a transition will be able to shift the status quo and challenge the structural drivers of exclusion and exploitation within the energy system (Swilling et al., 2016). Even when just transition discourses are actively mobilised by political elites, they seem to have a limited impact on the well-being and lives of those who are most disadvantaged (Johnston et al., 2023).

Environmental justice scholars have long linked decision-making power to the deployment of knowledge and the construction of specific forms of rationality (Patel, 2009). Postcolonial studies of transition show that pervasive forms of coloniality are reproduced in dominant narratives of economic development that underpin the injustices and inequalities that a just transition is deemed to address (Johnston et al., 2023). Coloniality is also reproduced in well-intentioned discourses of justice and well-being through the imposition of certain models of thinking, such as aspirations to rational or clean modernities (Tamale, 2020). Postcolonial scholars have called for action that acknowledges how imperial legacies influence today's unfair energy landscapes and thus shape the transition to sustainable energy (Swilling & Annecke, 2012).

In this context, community energy experiences in countries like Ethiopia, Malawi, and Mozambique reveal what future transitions are possible and how to deliver them. The CESET's team has adopted a broad definition of community energy, encompassing communities with a shared energy-related interest and deploying collective actions which democratise the energy sector and enrol citizens in the management and governance of energy systems (Rincón-Rubio & Cedano-Villavicencio, 2023). Central to this definition is the possibility of generating plural perspectives on energy and bottom-up-led actions that can challenge existing forms of coloniality embedded in current systems of energy provision (Routledge et al., 2018; Sovacool et al., 2023). Community energy is also a means to redefine the systems of knowledge that underpin transition thinking and policy prescriptions. In doing so the CESET experience extends well beyond these countries providing specific insights for the slow transition that will eventually build the cathedral of a sustainable future. Thus, the question is not only about the role of community energy in a transition to sustainable energy, but specifically, in which ways community energy can act as an engine for social change, and how does community energy counters existing inequities in energy provision and use, including the recognition of the forms of oppression and discrimination that shape the process and outcomes of transitions.

The innovation of this book is that it attempts to answer these questions from the perspective of the experience of existing community energy projects on the ground, attending to the different factors that shape their feasibility and their impacts on communities and on the politics of energy. This requires examining community energy from an interdisciplinary perspective. Community energy projects are diverse: they require diverse knowledge in terms of the multiple technologies, resources, governance arrangements, and sizes that shape community energy projects. They are also diverse in terms of their integrations of heterogeneous communities that come together working for a common interest, sometimes with variable results. These topics require interdisciplinary—sometimes uneasy—dialogue across engineering, economics, and critical social science, the kind of dialogue that CESET seeks to foster. The result is a book of equally diverse contributions, full of lessons learned and experiences, but also full of warnings about unduly optimism about the potential for community energy. The book calls for moving away from ready-made recipes for community energy in favour of co-designing projects from the bottom up, led by the concerns and skills within the communities deemed to benefit from these projects. The book is grounded on a propositional attitude to infrastructure development focused on future possibilities within specific contexts rather than a particularistic critique of the structural constraints that prevent fairer infrastructural futures (Baptista & Cirolia, 2022). Such a propositional attitude goes hand in hand with a recognition of the importance of provisional responses to imagine infrastructural alternatives.

In summary, this book presents an interdisciplinary, grounded, but forward-looking—cathedral-building—perspective on the transition to sustainable energy, using community energy as the main proposition to examine the process and consequences of such transition. The rest of the introduction discusses community energy as a pragmatic response to activate a bottom-up transition to sustainable energy first, then discusses how community energy relates to current debates on just transitions and energy justice. The final part of the introduction provides an overview of the book's contributions.

1.2 Community Energy as a Pragmatic Response to the Transition to Sustainable Energy

Community energy, sometimes also referred to as energy communities, generally refers to specific situations in which the provision and use of energy are governed locally. There is no single definition of community energy: a systematic review of community energy definitions showed that precise definitions tend to be developed in relation to policy problems or specific contexts of study (Brummer, 2018). At the same time, this review identified that, while not universal, community energy projects tend to:

- involve the deployment of renewable energy technology;
- rely on decentralised infrastructure networks;
- be developed at small, local scales;
- serving a local area where production and consumption of energy take place in close proximity;
- depend on diverse forms of participatory decision-making and community ownership.

Some scholars emphasise the notion of 'integration' (e.g., integrated community energy systems) in community energy because community energy projects generally involve a variety of local heat and electricity generation, flexible demand, and storage (Koirala et al., 2016). The development of online-based platforms has further enabled new models of peer-to-peer energy trading that have further diversified ideas about community energy and whether it is territorially located (Sousa et al.,

2019; Zhang et al., 2017). Definitions are built on stories, and stories are built on examples. There is perhaps no example of community energy most famous than the Samsø Renewable Island Project in Denmark, which achieved 100% renewable energy supply in the short span of a decade through collective and cooperative-based action (Sperling, 2017). Projects such as this have inspired community action in other locations around the world. However, the emphasis on regional, technologically savvy projects distracts attention away from the fact that the majority of community energy projects are initiated in contexts with scarce resources in which communities work together to access markets and technologies. While technological advancements enable innovation in community energy (e.g., Huang et al., 2015), there is a risk that complicated technologies and larger-scale infrastructures will exclude the communities that should be at the centre of the action.

A major challenge in understanding community energy is precisely explaining what 'community' is. The idea of community is often taken for granted, but the literature shows that communities are neither bound to place nor delimited social networks (Creamer et al., 2018). Instead, community refers to a dynamic social group which is actively constructed through the process of building a common project. This is explored in depth in Chapter 3.

The question of understanding 'community' relates to one of the main features of community energy, having community ownership. Community ownership is not simply a pragmatic arrangement to manage a project. Rather, community ownership is one of the central features whereby community energy may enable the democratisation of the energy system, facilitating users' participation in decision-making concerning energy services (Burke & Stephens, 2018). The challenge here is identifying who is the subject making decisions and how decisions are made. Communities are not homogeneous units with a single rationality (Rigon & Castán Broto, 2021). Moreover, this raises questions about whether there is a separation between the knowledge held by communities and the knowledge provided by energy experts, often from outside the community.

What is the contribution of community energy to transitions to sustainable energy? Community energy provides multiple benefits. Table 1.1 adapts the results of a systematic review, which examined the benefits of community energy documented in the US, Germany, and the UK (Brummer, 2018). The analysis is shaped by policy assumptions made in North American and European contexts, which may not be entirely relevant in Eastern and Southeastern African countries. Nevertheless, Table 1.1 shows the range of economic and social benefits associated with community energy, including building knowledge for future changes and developing endogenous innovation. At the same time, Table 1.1 demonstrates that the contribution of community energy to the transition is most often framed in terms of contributing additional generation capacity or in terms of raising awareness for changing lifestyles and improved attitudes towards renewables (e.g., Bauwens & Devine-Wright, 2018). Community energy is framed as an add-on to existing infrastructure configurations rather than a mechanism to change the fundamental technological regimes that shape them.

More recent work, however, has explored how community energy reconfigures infrastructure systems, in particular, through facilitating the decentralisation of energy generation. For example, a comparative study of 13 case studies of community energy in the Netherlands (including solar PV, biomass, and increased efficiency projects) investigated their contribution to the decentralisation of the energy network and the subsequent adoption of renewables (The Netherlands lags behind other European countries in the share of renewables within their energy mix) but were unable to reach a conclusion because the relatively young character of all the initiatives investigated (van der Schoor & Scholtens, 2015). Even in countries with a strong community energy policy, the spread of community energy remains aspirational.

A few factors may change the context of the development of community energy. First, as projects get slowly consolidated, there is growing evidence that community energy is a feasible response to providing energy access, particularly in contexts where there are no alternatives. The drive to attain Sustainable Development Goal 7—of providing universal energy access to clean energy by 2030—has revealed that current strategies for infrastructure development will not serve to bridge the enormous gaps in energy access, even more now that evidence increasingly shows that energy poverty is frequently reproduced under the grid (González-Eguino, 2015). Decentralised, modular, flexible systems are also considered more adaptable to changing conditions of infrastructure provision as rapid urbanisation and climate change deepen current infrastructure gaps. The benefits of community energy are thus not only related to economic benefits and the decarbonisation of the energy system but also to future

Table 1.1 Community energy benefits

Economic	Energy access Financial benefit for the community Services for marginalised areas or communities Higher employment Social inclusion Support of other community activities and services
Education and acceptance	Knowledge about energy-saving Understanding how to run community projects Examples that can inspire other communities Improving trust and acceptance towards renewable energy
Participation	Higher level of political participation Collective financial management Self-organisation
Climate protection and sustainability	Awareness and lifestyle changes Provision of low carbon supply of energy
Community building and self-realisation	Community upgrading Social cohesion and more robust governance Pride, joy, and other emotions related to collective material action
RE generation targets	Increase the share of renewables in the energy supply Level the playing field for market entrants Building supply chains to facilitate RE adoption
Innovation	Endogenous innovation Generation of new societal norms

Adapted from Brummer (2018)

visions that emphasise resilience as the main value shaping the design of infrastructure systems (see Chapter 2).

Community energy challenges the model of infrastructure provision that still dominates transition thinking. It proposes models for imagining infrastructure reconfiguration alongside the possibilities of more flexible technologies and away from networked models of urbanism and infrastructure planning (Coutard & Rutherford, 2015). Post-networked narratives may help assemble a different perspective away from the emphasis on infrastructure inadequacy and failure that pervade accounts of infrastructure development—particularly in Africa—and that seems to miss the embeddedness of infrastructures in systems of social and political integration (Silver, 2023). A post-networked model appears more attuned to the realities of infrastructure development in those locations where networked models of infrastructure never delivered on their promise of universal reach, but it is also associated with new patterns of inequality and displacement (Essex & de Groot, 2019). Because of its refusal to accept the traditional public and private-led models of energy provision, community energy poses a direct challenge to the institutional organisation and the actors of transitions, creating de facto new energy governance models of which the cooperative is but one example (Avelino & Wittmayer, 2016). For those who conceive the transition as a political struggle, the potential for community energy to change the fundamental structures of decision-making structures is their major asset (Burke & Stephens, 2018). Community energy operates both as an add-on that provides a Band-Aid solution to an inadequate energy system that perpetuates energy injustices and as a tool to potentially challenge the existing infrastructural politics of energy. Hence, it is imperative to consider community energy not just as a means for transitions but as a means for moving towards a just transition ideal.

1.3 Community Energy as an Instrument to Advance Justice in Transitions

CESET's foundational hypothesis is that community energy can contribute to making transitions to sustainable energy more just. A growing body of literature on energy justice informs this hypothesis. A certain consensus has emerged about defining energy justice in relation to the three pillars that have evolved with the literature on environmental justice: fair distribution of benefits and impacts, recognition of multiple subject positions and interests, and plural and active participation in decisions concerning energy provision and use (for some recent examples see: Jenkins et al., 2016, 2021). Within the transition, community energy can be seen to directly contribute to reduce the enormous inequalities in access to energy while also providing opportunities to foreground marginalised perspectives on energy, move beyond technocratic prescriptions for energy development, prioritising social, human-centred development, and enabling people to actively participate in the development of renewable technologies. These outcomes, however, are heavily dependent on the manner of implementation, and community energy initiatives are not inherently just. Moreover, putting justice at the centre of the transition does not actually translate into resolving the broad drivers of discrimination that result in an unequal energy system: justice perspectives raise dilemmas that may be overlooked in the context of urgency—such as, for example, the need to mobilise a community around a common purpose to deliver the project's benefits in a timely and efficient manner (Kumar et al., 2021).

Community energy projects are sometimes assessed in terms of their contribution to energy justice in ways that overlook the contradictions and dilemmas that emerge at the core of energy projects. However, the perspective on community energy projects changes when they are situated in broader contexts of energy production and consumption within the energy sector. Healy et al. (2019) have proposed to consider embodied energy injustices, that is, the interlinked chain of injustices that occur during the extraction, processing, transportation, and disposal of energy. This means that assessing how community energy addresses energy injustices cannot be limited to the specific moment of project conception and delivery. Inspired by this notion, van Bommel and Höffken (2021) have argued that community energy justice can be analysed at three different levels: actions within the initiatives themselves, in between the relations of those initiatives with a broader set of actors, and beyond the initiatives. This framework allows for a multidimensional understanding of justice within community energy, which reveals its interaction with the larger processes of social change that shape the transition to sustainable energy (Table 1.2).

Table 1.2 contains multiple lessons. The first one is that advancing justice requires multiple actions at many levels, from the practical, localised action that takes place within independent community energy projects to the forms of narrative and institutional change that take place in the interactions within the networks of stakeholders that shape them, and within the broader political system. It is this third dimension that becomes more fundamental for understanding transitions. Here, community energy reveals injustices and, through cumulative experiences, helps build alternative visions of sustainable futures. However, it would be an error to focus exclusively on these aspects of community energy development. For that reason, the focus of this book is not on discussing justice

Table 1.2 An expanding perspective of the role of community energy in advancing just transitions

Level of analysis to understand community energy justice	*Relational dynamics that enable just/unjust action*	*Examples of justice issues raised in relation to community energy*
Within	Community dynamics, decision-making, exclusion of vulnerable groups	Distribution of community energy benefits, changing the dynamics of community (e.g., new livelihood and education opportunities) Participation of the community and transparency in energy decision-making within community energy initiatives, with opportunities for involvement Recognition of differential needs within the community
Between	Interaction with multiple parties, including policymakers, intermediaries, developers	Shifting interests of external parties, for example, recognising new user groups and business opportunities Shifting of governmental responsibilities to facilitate alternatives to the dominant regime, for example, through off-grid regulations or community energy plans Development of funding programmes within government and international development organisations to support the democratisation of the energy system Reveal the limitations of erratic public policies, the prioritisation of large public or private monopolising companies, and the inadequacy of development programmes that do not cater to local community interests

Level of analysis to understand community energy justice	*Relational dynamics that enable just/unjust action*	*Examples of justice issues raised in relation to community energy*
Beyond	Community energy development within an energy system in transition	Claim for opportunities to initiate community energy, making the technology accessible to wider sectors of the population and helping democratise energy supply Challenging embodied energy injustices by claiming autonomy and energy sovereignty Demonstrating the cumulative power of multiple community projects challenging the existing energy system

Expanding the analysis by van Bommel and Höffken (2021)

within projects but on understanding how community projects unfold within specific political contexts, activating changes across different levels of interaction.

1.4 Structure of the Book

This book focuses on bringing in perspectives from the countries of Ethiopia, Malawi, and Mozambique. These are all countries where there is a dynamic discussion on community energy and what it means for the development of the energy sector in each country. The discussions in these countries resonate with ongoing efforts to conceptualise the energy transition in the African context (Sokona et al., 2023). In addition, the discussions of the contribution of community energy to just transitions in those countries contain powerful lessons about the theory and practice of community energy that can be valuable elsewhere as community energy becomes a demanded alternative for sustainable energy. This contribution is even more critical in the current research context, dominated by anglophone perspectives on community energy that present as universal very provincial debates. For example, a recent bibliographic review analysed 263 articles on community energy from 1997 to 2018 and found that only 2% of the articles had a geographical location outside Europe and North America (van der Schoor & Scholtens, 2019). In contrast, 36% of the cases were in the UK and 10% in the US. Thus, empirical research from the UK and the US has driven interest in a body of research focused on how community energy impacts the acceptance of renewables—an issue that relates to the dynamics of the setting of renewable energy facilities in those countries, but that would be framed differently in other locations. Even the discussion provided in this chapter, based on a phronetic appraisal of the literature, rather than on a systematic review, demonstrates the dominance of the UK experience in current thinking on community energy. In this context, the empirical experiences of other countries where community energy is developing rapidly or changing transition discourses and the theoretical insights from those processes may be particularly important to challenge context-based assumptions that dominate current models and recommendations for community energy.

The CESET team is committed to developing a theoretical perspective on community energy from elsewhere. The contributions to this book are not only characterised by coming from situated locations (in relation to critical analyses of energy systems in Ethiopia, Malawi, and Mozambique), but also from multiple disciplinary outlooks—because the CESET team is a recognition that energy transitions require an interdisciplinary effort to understand technological change alongside the massive changes in society and politics that would make such technological changes appropriate and sustainable. The book is roughly divided into two sections: the first section focuses on developing a conceptual toolbox that can bridge disciplinary perspectives in thinking policy and action for community energy. They are led by scholars actively engaged in the practice of community energy. The book's second part focuses on distilling practical lessons from engaged efforts on the ground led by engineers and practitioners of community energy.

The book's first half engages with five key concepts that challenge current assumptions about community energy. These are resilience, community, gender, finance, and regulation. Chapter 2 analyses the relationship between community energy and community resilience. The chapter challenges current approaches to the benefits of community energy. In diverse fields such as ecology and engineering, resilience refers to the capacity of a system to withstand shocks. This term has travelled to community development policy to highlight the sustainability of communities, but by emphasising permanence, the term has often been prescriptive of conservative action directed towards keeping communities stable or operating within a given system. This chapter challenges such notions by reframing resilience as a means to transform communities to emphasise social development and community well-being—building the different forms of capital that enable a thriving community. Vallecha and To offer a persuasive account of the multiple ways in which community energy enables multi-level transformations that ultimately may lead to the activation of resilient communities within a fundamental unjust energy system.

Chapter 3 complicates the narrative of community energy by examining the notion of community closely, particularly seeking to put aside myths such as the purported homogeneity of communities and the idea that communities can be found ready to act in specific locations. Instead, the chapter portraits communities as dynamic and purposeful social arrangements that come together to claim collective spaces within an area of intervention. The chapter develops a notion of community that, without claiming community utopias, can deliver pragmatic means to challenge injustice from the local (e.g., through the development of appropriate conflict management plans) to the global (e.g., by collectively generating assessments of embedded energy injustices).

Chapter 4 engages with a feminist analysis of energy justice as a means to put justice and inequality at the centre of the community energy debate, in line with the concerns about just transitions explained above. Such a feminist lens cannot be advanced without paying tribute to a long-standing tradition of gender scholars working in energy studies, whose contributions were often ignored in a highly masculinist context of research. However, some of the debates on gender and energy have sometimes tended to simplify the nature of gender relations and essentialise gender roles, leading to simplistic assumptions that generate inadequate policy. For example, improved cookstove programmes directed at women may have overlooked the complex relations around cooking while also preventing the development of alternatives such as electric cooking. The challenge is that we hardly understand how gender relations impact energy access and resilience within specific communities because much of the empirical work in this area falls back on essentialist stereotypes and lacks disaggregated data. The chapter proposes an intersectional perspective moving beyond those contexts of energy development.

Finance is the main subject of Chapter 5. From the outset, CESET aimed to advance new ideas on alternative forms of finance for community energy, and agitated debates took place within the project team. Aran Eales kindly took these debates forward by examining the team's lofty aspirations in relation to the actual constraints in drawing the capital and facilitating the management of community energy projects. Two key insights emerge from Chapter 5. First, while ideals of self-financing of community energy are persistent, in reality, most community energy projects rely on external grants for survival. In Malawi, for example, ODA-related financing plays a significant role in making community energy possible, but ODA funding is not always available. The possibility of developing forms of alternative finance is related to the increasing availability of cheap technologies for community energy. Second, capital is not sufficient to make community energy available: key bottlenecks appear once projects are up and running during its maintenance and management, but these costs are routinely underestimated. Communities have the tools to reduce and manage these costs, particularly when they are given support to develop appropriate skills.

Chapter 6 engages with the other sacred cow of community energy: regulation. This multi-institutional collaboration emerges from a collective dialogue on the common assumption that more regulation facilitates transitions to sustainable energy and energy access and, even more, the proliferation of community energy. An empirical analysis of statistical data in the three countries demonstrates that this relationship is uneven and that the relationship between the development of regulations and energy access strongly depends on contextual factors about the needs of the country and the ways in which regulations are implemented. Nevertheless, the chapter provides a compelling case for increasing attention to regulation to foster a transition to sustainability with universal reach.

Chapters 7–11 seek to mobilise some of these concepts in different regulation, education, and practice contexts in Ethiopia, Malawi, and Mozambique. Chapters 7 and 8 explore the context of Ethiopia, a country that has pioneered government-led efforts to deliver decentralised energy but whose ambitious energy plans were curtailed by the civil war in Tigray (2020–2022). The chapters are led by leading scholars in the country who, during the duration of CESET, have been involved in maintaining precarious energy infrastructures during the conflict. While Chapter 7 delivers a systematic analysis of the barriers to community energy within existing institutions and regulations in the country, Chapter 8 explores future potential through the development of education and skills needed for the energy transition. Chapter 8, in particular, incorporates some of the concerns about interdisciplinarity and gender developed in previous chapters to challenge the relatively homogeneous and technology-oriented landscape of energy education in the country. Chapters 9 and 10 focus on Malawi, a country with strong policies for community energy development fostered under the umbrella of international development programmes such as those led by the United Nations Development Programme (UNDP). Chapter 9, led by a research group at the University of Mzuzu, a leading international research centre on people-centred energy studies. The chapter analyses the uneasy history of community development in Malawi, from the initial localised projects that were developed with grant funding in the 2000s to the more coordinated programmes that are taking place today. The comparative experience of multiple projects provides practical insights about what matters for community energy development. Two case studies presented in Chapter 10 examine these lessons. In Chapter 10 two community energy practitioners explain their experiences seeking to inspire other practitioners working in the area. Chapter 11 brings together a group of experts on the political economy of energy systems in Mozambique to explain their perspectives on a changing regulatory context where top-down policies drive off-grid network development. The chapter shows the interrelated set of concerns that emerge during the implementation of regulatory frameworks as multiple actors and technologies interact in a changing landscape.

The final chapter comes back to the two main questions raised by the book about the role of community energy in fostering transitions and the extent to which community energy advances justice within changing contexts of energy provision and use. The chapter concludes that community energy holds significant promise but that this promise is invariably linked to its diversity: the diversity of technologies that characterise community energy and the diversity of people and social relations that characterise the communities that make it possible. Such diversity requires flexible approaches to developing community energy projects, seeking to take maximum advantage of the contextual opportunities offered by different political economy contexts but realising the inherent contradictions that community energy projects, always projects-in-the-making, may rise.

References

Avelino, F., & Wittmayer, J. M. (2016). Shifting power relations in sustainability transitions: A multi-actor perspective. *Journal of Environmental Policy & Planning, 18*(5), 628–649.

Baptista, I., & Cirolia, L. R. (2022). From problematisation to propositionality: Advancing southern urban infrastructure debates. *Transactions of the Institute of British Geographers, 47*(4), 927–939.

Bauwens, T., & Devine-Wright, P. (2018). Positive energies? An empirical study of community energy participation and attitudes to renewable energy. *Energy Policy, 118*, 612–625.

Brummer, V. (2018). Community energy—Benefits and barriers: A comparative literature review of community energy in the UK, Germany and the USA, the benefits it provides for society and the barriers it faces. *Renewable and Sustainable Energy Reviews, 94*, 187–196.

Burke, M. J., & Stephens, J. C. (2018). Political power and renewable energy futures: A critical review. *Energy Research & Social Science, 35*, 78–93.

Coutard, O., & Rutherford, J. (2015). *Beyond the networked city: Infrastructure reconfigurations and urban change in the North and South*. Routledge.

Creamer, E., Eadson, W., van Veelen, B., Pinker, A., Tingey, M., Braunholtz-Speight, T., Markantoni, M., Foden, M., & Lacey-Barnacle, M. (2018). Community energy: Entanglements of community, state, and private sector. *Geography Compass, 12*(7), e12378.

Essex, S., & de Groot, J. (2019). Understanding energy transitions: The changing versions of the modern infrastructure ideal and the 'energy underclass' in South Africa, 1860–2019. *Energy Policy, 133*, 110937.

González-Eguino, M. (2015). Energy poverty: An overview. *Renewable and Sustainable Energy Reviews, 47*, 377–385.

Healy, N., Stephens, J. C., & Malin, S. A. (2019). Embodied energy injustices: Unveiling and politicizing the transboundary harms of fossil fuel extractivism and fossil fuel supply chains. *Energy Research & Social Science, 48*, 219–234.

Heffron, R. J., & Heffron, R. J. (2021). What is the "just transition"? In *Achieving a just transition to a low-carbon economy* (pp. 9–19). Springer.

Huang, Z., Yu, H., Peng, Z., & Zhao, M. (2015). Methods and tools for community energy planning: A review. *Renewable and Sustainable Energy Reviews, 42*, 1335–1348.

Jenkins, K., McCauley, D., Heffron, R., Stephan, H., & Rehner, R. (2016). Energy justice: A conceptual review. *Energy Research & Social Science, 11*, 174–182.

Jenkins, K. E., Sovacool, B. K., Mouter, N., Hacking, N., Burns, K., & McCauley, D. (2021). The methodologies, geographies, and technologies of energy justice: A systematic and comprehensive review. *Environmental Research Letters, 16*(4), 043009.

Johnston, M., Darkey, D., & Ibsen, H. (2023). Environmental justice and dissent for postcolonial urban sustainability transitions. *International Journal of Urban and Regional Research, 47*(4), 645–664.

Koirala, B. P., Koliou, E., Friege, J., Hakvoort, R. A., & Herder, P. M. (2016). Energetic communities for community energy: A review of key issues and trends shaping integrated community energy systems. *Renewable and Sustainable Energy Reviews, 56*, 722–744.

Kumar, A., Höffken, J. I., & Pols, A. (2021). *Dilemmas of energy transitions in the global south: Balancing urgency and justice*. Taylor & Francis.

Lee, H., Calvin, K., Dasgupta, D., Krinner, G., Mukherji, A., Thorne, P., Trisos, J. Romero, P. Aldunce, Barret, K., & Park, Y. (2023). *IPCC, 2023: Climate change 2023: Synthesis report, Summary for policymakers. Contribution of working groups I, II and III to the sixth assessment report of the Intergovernmental Panel on Climate Change* (Core Writing Team, H. Lee & J. Romero, Eds.). IPCC.

Newell, P., & Mulvaney, D. (2013). The political economy of the 'just transition.' *The Geographical Journal, 179*(2), 132–140.

Novalia, W., Rogers, B. C., & Bos, J. J. (2021). Incumbency and political compromises: Opportunity or threat to sustainability transitions? *Environmental Innovation and Societal Transitions, 40*, 680–698.

Patel, Z. (2009). Environmental justice in South Africa: Tools and trade-offs. *Social Dynamics, 35*(1), 94–110.

Rigon, A., & Castán Broto, V. (2021). *Inclusive urban development in the global south: Intersectionalities, inequities, and community*. Routledge.

Rincón-Rubio, A. G., & Cedano-Villavicencio, K. G. (2023). Emotional energy communities: Centering emotions and feelings within energy transitions in southern Mexico. *Energy Research & Social Science, 98*, 103014.

Routledge, P., Cumbers, A., & Derickson, K. D. (2018). States of just transition: Realising climate justice through and against the state. *Geoforum, 88*, 78–86.

Silver, J. (2023). *The infrastructural south: Techno-environments of the third wave of urbanization*. MIT Press.

Sokona, Y., et al. (2023). *Just transition: A climate, energy and development vision for Africa*. Independent Expert Group on Just Transition and Development. Retrieved January 19, 2024, from https://justtransitionafrica.org/wp-content/uploads/2023/05/Just-Transition-Africa-report-ENG_single-pages.pdf

Solnit, R. (2024). *Slow change can be radical change*. Retrieved January 19, 2024, from https://lithub.com/rebecca-solnit-slow-change-can-be-radical-change/

Sousa, T., Soares, T., Pinson, P., Moret, F., Baroche, T., & Sorin, E. (2019). Peer-to-peer and community-based markets: A comprehensive review. *Renewable and Sustainable Energy Reviews, 104*, 367–378.

Sovacool, B. K., Bell, S. E., Daggett, C., Labuski, C., Lennon, M., Naylor, L., Klinger, J., Leonard, K., & Firestone, J. (2023). Pluralizing energy justice: Incorporating feminist, anti-racist, Indigenous, and postcolonial perspectives. *Energy Research & Social Science, 97*, 102996.

Sperling, K. (2017). How does a pioneer community energy project succeed in practice? The case of the Samsø Renewable Energy Island. *Renewable and Sustainable Energy Reviews, 71*, 884–897.

Swilling, M., & Annecke, E. (2012). *Just transitions: Explorations of sustainability in an unfair world*. Juta and Company (Pty) Ltd.

Swilling, M., Musango, J., & Wakeford, J. (2016). Developmental states and sustainability transitions: Prospects of a just transition in South Africa. *Journal of Environmental Policy & Planning, 18*(5), 650–672.

Swyngedouw, E. (1997). Power, nature, and the city. The conquest of water and the political ecology of urbanization in Guayaquil, Ecuador: 1880–1990. *Environment and Planning A, 29*(2), 311–332.

Tamale, S. (2020). *Decolonization and Afro-feminism*. Daraja Press.

Thunberg, G. (2019, April 23). You did not act in time. *The Guardian*. https://www.theguardian.com/environment/2019/apr/23/greta-thunberg-full-speech-to-mps-you-did-not-act-in-time

van Bommel, N., & Höffken, J. I. (2021). Energy justice within, between and beyond European community energy initiatives: A review. *Energy Research & Social Science, 79*, 102157.
van der Schoor, T., & Scholtens, B. (2015). Power to the people: Local community initiatives and the transition to sustainable energy. *Renewable and Sustainable Energy Reviews, 43*, 666–675.
van der Schoor, T., & Scholtens, B. (2019). The power of friends and neighbors: A review of community energy research. *Current Opinion in Environmental Sustainability, 39*, 71–80.
Wang, X., & Lo, K. (2021). Just transition: A conceptual review. *Energy Research & Social Science, 82*, 102291.
Zhang, C., Wu, J., Long, C., & Cheng, M. (2017). Review of existing peer-to-peer energy trading projects. *Energy Procedia, 105*, 2563–2568.
Zografos, C. (2022). The contradictions of Green New Deals: Green sacrifice and colonialism. *Soundings, 80*(80), 37–50.

CHAPTER 2

Community Energy and Community Resilience: A Multi-Dimensional Perspective

Harshit Vallecha and Long Seng To

Abstract This chapter explores the mutually interdependent relationship between community energy and community resilience. The resilience of Community Energy Systems (CES) depends on both the physical aspects of energy provision and the socio-economic and political aspects of the constitution of communities. At the same time, sustainable energy projects managed by communities build autonomy, control over resources, and community cohesion thus making communities themselves more resilient. This chapter offers an analytical framework to analyse resilience in CES and to increase their resilience during their conceptualisation, installation, operation, and upgradation. Resilience is seen here as dependent on communities' social, economic, physical, and human capital. The framework shows that while some aspects of CES resilience

H. Vallecha (✉) · L. S. To
STEER Centre, Geography and Environment, School of Social Sciences and Humanities, Loughborough University, Loughborough, United Kingdom
e-mail: H.Vallecha@lboro.ac.uk

L. S. To
e-mail: L.To@lboro.ac.uk

V. Castán Broto (ed.), *Community Energy and Sustainable Energy Transitions*, https://doi.org/10.1007/978-3-031-57938-7_2

can be addressed in the design of the energy system, their success depends on appropriate mechanisms for community involvement and governance.

Keywords Community energy · Community resilience · Human capital

2.1 Introduction

Community resilience is now a concern for many local governments and neighbourhood institutions that provide essential services to people living in cities and settlements. Resilience is the capacity or ability to avoid, withstand, and recover quickly from any unforeseen disruption or shock. Resilience is often used concerning systems, that is, complex assemblages of elements that operate as a whole. When resilience is defined as a characteristic of a system, it emphasises its context and operation. Energy services are pivotal for any community; thus, they are central to community resilience.

A community often refers to a group of people who recognise a shared identity or interest, often associated with a given location (see Chapter 3). In a local context where a group of people call themselves 'community', community resilience refers to the ability of such a community to cope with hazardous events or disruptions. For example, communities may respond or reorganise to maintain their integrity, functions, identity, and structure while maintaining capacities for adaptation, learning, and transformation (Cretney, 2014). Community resilience relates to individuals' expectations and actions, to collective operations and institutions, and to the material conditions they live in. Community Energy (CE) projects may help build community resilience, but they also depend on maintaining community integrity; thus, community energy and community resilience are closely interlinked.

In Community Energy Systems (CES), resilience relates to both the physical aspects of energy provision and the socio-economic and political aspects of the constitution of communities. The capacity of CESs to advance community resilience depends on the material and infrastructural conditions of the system and whether they can endure unpredictable events such as a sudden surge in demand, unexpected meteorological events, or major system failures. However, that capacity also depends on the community's characteristics, including the economic and social capital of the people in the community, the institutions that regulate it, and the human reaction to the events that test the system. Physical systems alone cannot endure changes without mobilising human adaptive capacities to deal with those changes.

Another challenge relates to the fact that the degree of participation of communities in community energy projects varies depending on the design of the project and the different stages of project implementation. Community involvement in the planning, installation, and operation phases of a community energy project builds the community's economic, social, physical, and human capital to cope with any risks or disruptions. The development of institutions and energy management practices in community energy projects also enables the community to manage the system during unforeseen events. For example, constructing a community-owned solar microgrid may bring the community together to develop a wide range of community-based projects, build a structure to facilitate leadership, and enable learning about energy infrastructures so that people can react appropriately when they fail. In summary, community energy develops community resilience through the physical construction of infrastructures, the development of management institutions at the community level (which help to maintain the community's identity and structure), and the collective capacity to develop adaptive reactions to unforeseen events.

Resilience is crucial in locations prone to natural hazards like earthquakes, cyclones, or wildfires. With the continuous rise in global emissions and climate change, such hazards are becoming more frequent across risk-prone regions. Every year, millions of people are displaced or affected by events whose increase in frequency and severity can be linked to climate change. Many of these events impact people's lives directly by destroying life-saving infrastructures. They also have cascading impacts which extend over larger regions, for example, when flooding causes a large-scale blackout. For instance, in 2019, cyclone Idai severely affected parts of Mozambique, Zimbabwe, and Malawi, causing strong winds and flooding that killed 1,593 people and destroyed over 100,000 homes. Idai also had an extended impact through the damages to the electricity network and the subsequent cholera epidemics, with 3 million people eventually impacted (Disasters Emergency Committee, 2019). In events like this, it may take several months to repair the damage and restore the power system to its normal state; thus, community energy provides respite.

Communities living near coastal, hilly, and national borders may face additional governance challenges when exposed to extreme events and socio-political conflicts. Due to their remote geography, they may lack access to national grid infrastructure. The deployment of CES in such off-grid communities provides reliable electricity access and makes them resilient to unforeseen risks or disruptions through evolved community capacity. Communities can make decisions in real life, protecting the energy supply and facilitating access to electricity in cases of shock.

Some of the past studies discussed the notion of resilience across remote and displacement settings. However, few studies have explored community energy systems' interactions with community resilience against unforeseen disruptions. This chapter discusses the inter-relationship between CES and resilience by explaining how CES contributes to building community's resilience and how resilience determines the long-term sustainability of a CE project. It describes the core stages of disaster management and the community's capacity to respond in each step. It further explains the influence of a community's resilience attributes towards the implementation cycle of a CE project and the core aspects of building a resilient energy community.

2.2 What Is Community Resilience?

Community resilience is the ability of a community to respond to crises while maintaining its integrity and strengthening its capacity to cope. As defined by the Community and Regional Resilience Institute (CARRI), "Community Resilience is the capability to anticipate risk, limit impact and bounce back rapidly through survival, adaptability, evolution and growth in the face of turbulent change" (Community and Regional Resilience Institute, 2013). Norris et al. (2008) define community resilience against unforeseen disruptions with respect to five components: 'shock', 'capacity', 'impact', 'trajectory', and 'outcomes'. Community resilience and its response cycle to an unforeseen event is further represented in Fig. 2.1.

'Shock' is a possible unforeseen event, trend, or disruption that disturbs the community. Events may have various causes, from disasters to conflict-related events such as invasions, pandemics, infrastructure breakdowns such as power grid collapse, and economic crises (Coaffee & Lee, 2016). In every case, the severity of the impacts of the shock will depend on the interaction of the shock with the conditions in which it

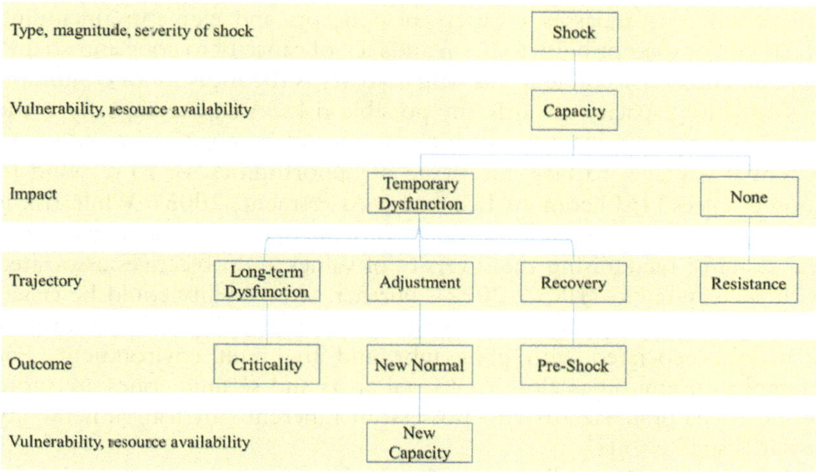

Fig. 2.1 Community resilience to an unforeseen event (adapted from Dabson, 2015)

impacts populations and assets. Tierney (2015) classified shocks under three categories—emergencies, disasters, and catastrophes.

Emergencies are classified as first-degree events with comparatively low severity, which could be managed internally without seeking external support. Disasters are defined as events with high severity which cannot be managed by local authorities and require external support like interventions from government or state agencies. For instance, the Nepal earthquake in 2015 was a severe disaster that killed over 9,000 people and displaced millions (To & Subedi, 2019). The Tohoku earthquake and Tsunami in Japan claimed more than 15k casualties while displacing nearly 450k people (National Geographic Society, 2023). Catastrophes are extreme events which affect the lives of masses across the countries and require support from national and international levels. One example of catastrophes is the 2004 Tsunami, which claimed the lives of more than 225k people (World Health Organization, 2023).

The second component, 'capacity', refers to the availability of resources and the community's inherent vulnerability towards a possible threat or unforeseen event. As per IPCC (Intergovernmental Panel on Climate Change), "Vulnerability is the propensity or predisposition to be adversely affected. It encompasses a variety of concepts and elements, including sensitivity or susceptibility to harm and lack of capacity to cope and adapt" (IPCC, 2022). In simple terms, vulnerability is the measure of community's adaptive capacity towards any possible risks. Adaptive capacity is the "ability of systems, institutions, humans and other organisms to adjust to potential damage, to take advantage of opportunities, or to respond to consequences" (Millennium Ecosystem Assessment, 2005). While risk is defined as the "potential for adverse consequences for human or ecological systems, recognising the diversity of values and objectives associated with such systems" (IPCC, 2022). Further, vulnerability could be classified as physical, economic, and social vulnerabilities. Physical vulnerability is mainly associated with geography and the built environment. For example, communities close to coastal areas and seismic zones are more prone to natural hazards and possess an inherent physical vulnerability towards such events.

Economic vulnerability depends on the economic structure of the community. Communities, where people depend on one single means of livelihood suffer more after disruption than the community with diversified livelihoods. Communities with weak economic structures demonstrate slow recovery after disturbance and become more vulnerable to financial crisis post-disruption. Conversely, social vulnerability is associated with inequalities across the community. A community where all the members do not have equal access to resources is more prone to risks than a community where all the members have equitable access to resources. Social inequality or discrimination, often based on identity variables (gender, age, caste, race, etc.), shapes access to resources and decision-making. Social inequality thus prevents certain sections of society from absorbing the severities resulting from an unforeseen event and limits them from accessing relief/recovery resources during or after a disruption.

The third component, 'impact', measures the intensity of shock concerning the community's inherent capacity to absorb it. If the community shows sufficient resistance to disruption, it will mitigate any adverse effects and continue operating normally. However, if the community fails to resist the severity of disruption due to insufficient capacity, it will result in temporary dysfunction. The disruption will have a severe effect, and daily life in the community may cease temporarily. The impact of the disruption will shape the path or 'trajectory' followed by the community in due course. If the community manages to overcome the phase of temporary dysfunction, it may recover and restore the community to pre-shock conditions. Commentators highlight, however, that these disruptions and impacts should be used to address structural conditions of vulnerability and thus return to an improved social and economic situation. The question is, however, not whether the shock may support social change but whether the community can recover from a permanent long-term dysfunction.

Another possibility is the adoption of a trajectory of adjustment, where the community shows the exceptional capacity to adapt against any adversities and reaches a state of 'new normal' with enhanced capacity to avoid further disruptions. It could be further described through the resilience recovery loss curve (Fig. 2.2), explaining how an acute disturbance changes the path of a community's functional capacity at one instance.

Figure 2.2 shows that communities with sufficient inherent capacity to resist the shock follow path B; after experiencing some social and economic loss, they recover to their original state. However, less resilient communities follow path C, where they experience additional socio-economic losses and recover to a capacity lower than the pre-shock level. On the other hand, more resilient communities follow path A, where after experiencing some social and economic loss, they recover to a capacity higher than the pre-shock level. Such communities return to a 'new normal' while adapting towards any adversities with enhanced capacity to avoid any further disruptions.

2.3 Assessing Community Resilience

The resilience of a community can be assessed based on the resources or assets available within the community and its response towards disruption management phases. There are typically four disruption management phases, 'Preparation', 'Absorption', 'Recovery', and 'Adaptation', when responding to an unforeseen event (Fig. 2.3). 'Preparation' refers to planning any pre-disruption activity or action to protect any loss of life and property during disruption. It includes predicting adverse events, training residents for emergency response, developing backup capacity, and other proactive tasks (Mayunga & Peacock, 2010). One possible example of preparedness could be creating an alternative water channel for a micro-hydro-based CE project to manage the generation crisis in the event of

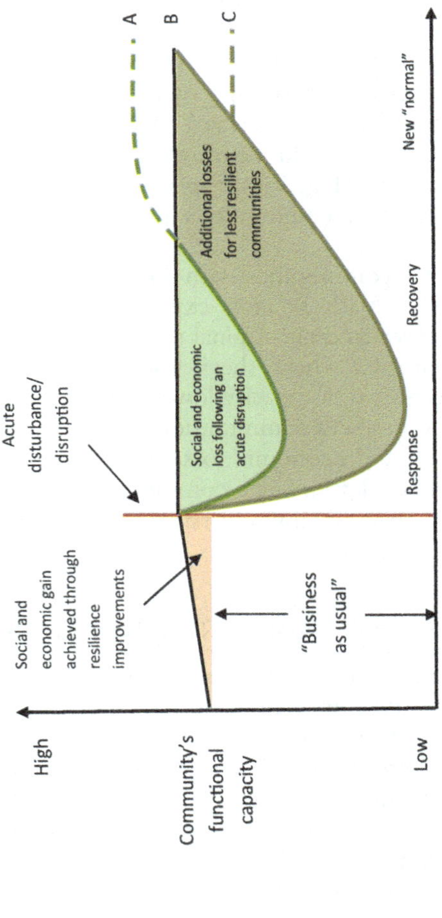

Fig. 2.2 Resilience recovery loss curve (adapted from White et al., 2014)

discontinuity in the water supply due to any natural obstruction like a landslide.

'Absorption' is the phase to reduce the impact of disruption while making critical services functional during an unforeseen event. For instance, disconnecting some commercial shops and businesses to manage power scarcity if a cyclone hits a solar-based CE project while maintaining continuous supply for essential services and domestic consumers. 'Recovery' refers to taking actions to restore the community's functions by keeping all the assets, services, and stakeholders functional to the same level as it was in the pre-disruption state. It includes recovering any loss to machinery and workforce by repairing or replacing faulty equipment

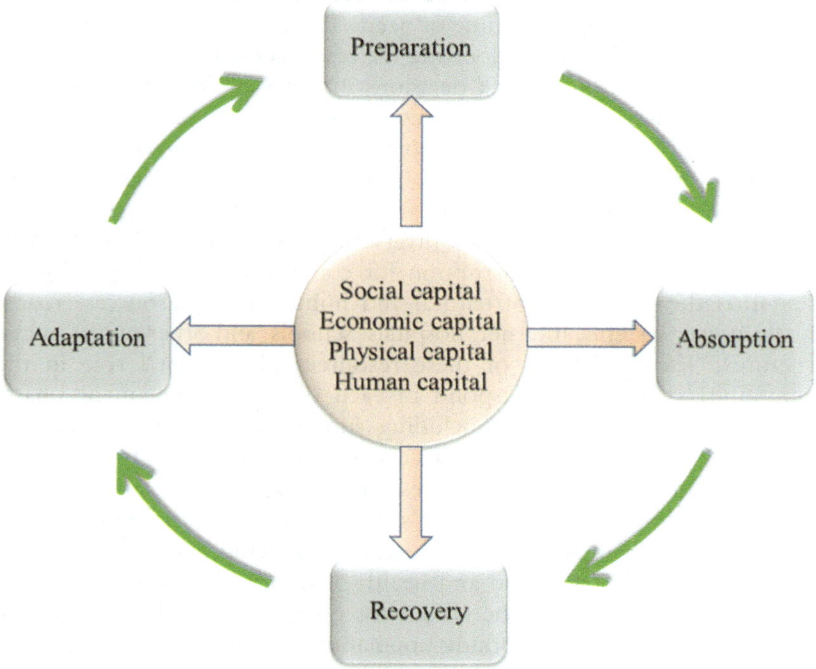

Fig. 2.3 Disruption management phases and community resilience (*Source* Author's own)

and training new operators in case of damage to a typical CE project (Oloruntoba et al., 2018).

'Adaptation' means learning from previous events and introducing changes in the original structure to mitigate further hazards with enhanced resiliency. It implies analysing the cause of failure or damage, upgrading machinery or tools, and designing adaptive policies and frameworks to reduce and possibly avoid any damage from further disruptions (Kuhlicke et al., 2023).

Conversely, community assets or capital can be broadly classified as Social, Economic, Physical, and Human capital. 'Social capital' measures the community's connectedness, trust, and cooperation among members to facilitate collective action during disruption. It depends on the social structure, norms, and bonds among the people to utilise their network of friends and relatives in their time of need. The community's social capital proved to be instrumental in locating the resources within and outside the community and making quick decisions in times of crisis (International Federation of Red Cross & Red Crescent Societies, 2014).

'Economic capital' is the availability of financial resources and the strength of the local economy. Financial resources like disaster recovery funds, insurance, access to credit, and state transfers are some features of sound economic capital. Also, diverse income streams will likely facilitate quick recovery after disruption for community members. 'Physical capital' is the availability of natural and human-made resources (roads, canals, power systems, internet, hospitals, schools, and other public infrastructure). The availability of these resources plays a vital role in the community's capacity to respond in each disaster management phase. Access to natural resources, including sun, wind, water, and biomass, may play a role as significant as access to physical capital (International Federation of Red Cross & Red Crescent Societies, 2004).

'Human capital' is the workforce's ability that enables a community to quickly recover from an economic crisis by producing goods and services. It is related to the education and health of the labour force in a typical community. Education of the labour force implies a level of awareness, skills, and training of the working population, which makes a community self-reliant to devise solutions and strategies to overcome disruption. The health of the labour force is equally essential, as the unhealthy working population cannot efficiently utilise other forms of community capital (Peacock et al., 2010).

A community's capacity in the form of social, economic, human, and physical capital measures a community's resilience towards an unforeseen event or disruption and the degree to which the community can prepare by anticipating potential risks or disruptions and adopting measures to minimise the severity from a potential disruption (Fig. 2.3).

2.4 How Do Community Energy Systems Contribute to Building Community Resilience?

Since community energy initiatives involve local community participation in one or more of the project implementation phases, deploying a CES develops community resilience across multiple dimensions. Implementing a CE project involves three core stages—conceptualisation or planning, installation, and operation. The involvement of the local community in one or all of the stages contributes to building the community's capacity in the form of the community's social, economic, physical, and human capital. Deploying CE projects builds the community's capital and develops awareness of using community assets to respond to unforeseen events. Figure 2.4 represents how an energy-poor community lacking reliable and sustainable energy sources was transformed into a new socio-economic capacity after the deployment of a CE project. It explains how the CE project implementation cycle involves the community's shared vision of innovation, need assessment, awareness, and motivation.

It further describes the formation of local alliances among community members for capacity building and developing problem-solving and learning attitudes. All these traits contribute to building the community's capital and make it resilient to respond against any disruptions. For instance, planning and installing a CE project develops human capital as the community becomes involved in technical learning and training activities. This specialised knowledge and approach for planning a socio-technical project while considering multi-dimensional factors build the community's capacity to devise strategies to prepare for an emergency.

The management of a CE project is usually performed by an institutional arrangement that operates as an 'Energy Committee', formed by elected members of the community. Members of energy committees are generally responsible for setting tariffs, fixing consumption, planning, maintenance, and upgrading. Forming these energy committees and volunteering groups helps build social capital by developing

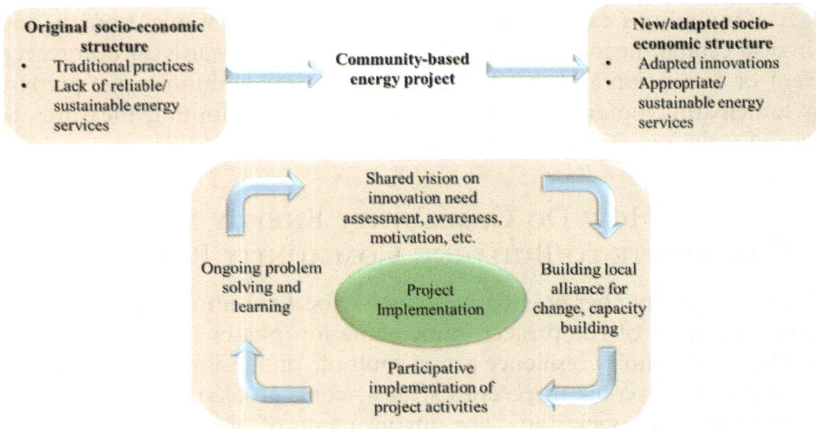

Fig. 2.4 Community transformation through a CE project (adapted from Ortiz et al., 2012)

traits of connectedness, cooperation, and local decision-making, further supporting all the phases of disruption management.

Deploying a CE project builds the community's physical capital by establishing electricity infrastructure and enabling communication facilities. These assets are believed to be the lifelines while recovering from an unforeseen event. Also, establishing a CE project generates employment opportunities, facilitates local businesses, and contributes to building the community's economic capital and making it resilient to recover from any financial crisis.

The community resilience aspects developed by the formation of community capital during the deployment of a CE project contribute to different phases of a CE project lifecycle. For instance, the social and human capital developed during the planning of a CE project will, in turn, contribute to the project's installation and operation, as shown in Fig. 2.5. Similarly, physical and economic capital developed during the installation and operation of a project will contribute to the upgrade phase. It implies that the community's capacity developed through each project implementation phase contributes to subsequent development. In

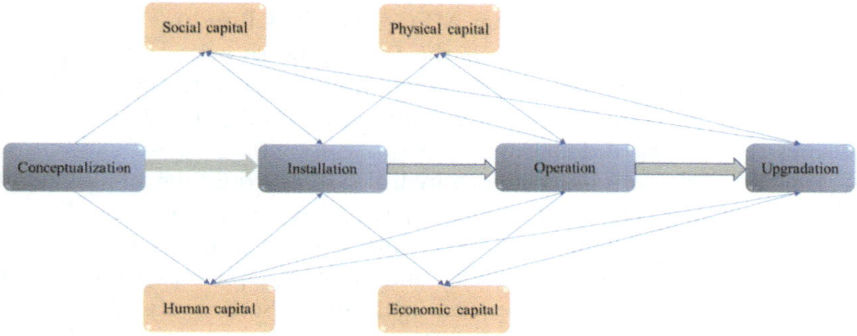

Fig. 2.5 CE project lifecycle and community resilience (*Source* Author's own)

other words, attributes of resilience evolved during various project implementation phases contribute to the efficient planning and execution of a CE project, making it sustainable in the long run.

2.4.1 *The Case of Bondo Micro-Hydro Community Energy System, Malawi*

Bondo is a small village in the Mulanje district in the southern part of Malawi. Mulanje is a mountainous region with several hills and perennial water streams. It is home to Mount Mulanje, the highest mountain peak in Malawi. Village Bondo in Mulanje was not electrified through the national grid due to Malawi's lack of electrification infrastructure. With its hilly terrain and remote geography, it was hard to realise the extension of the grid in the near future by residents of Bondo. Thus, a local conservation body, Mulanje Mountain Conservation Trust (MMCT), took the initiative to develop a micro-hydro-based CE project by utilising natural perennial streams of water for the people of Bondo. With the help of funding available through some international organisations and local community support, MMCT started the installation of Bondo-1 micro-hydro powerhouse in 2012 and established a mini-grid to extend electricity connections to the villagers. Later, Mulanje Energy Generation Agency (MEGA) was established in 2014, wholly owned by MMCT. MEGA is a social enterprise and Malawi's first licensed Independent Power Producer. It was set for sustainable operation and expansion of the Bondo micro-hydro scheme. MEGA installed Bondo-2 and Bondo-3 powerhouses and synchronised the three powerhouses to feed power in one mini-grid. Presently, three powerhouses generate ~268 kW and provide electricity to more than 2000 households, along with some businesses, schools, and charities.

The Bondo community micro-hydro system could be an excellent example of how the community's capacity has evolved during the CE project implementation phases. Deployment of the Bondo-1 powerhouse developed the community's social and human capital as local people were involved during planning and installation. Some were trained and acquired essential skills to carry out routine maintenance. The deployment of a powerhouse strengthened the social fabric among community members, and establishing an energy committee enabled decision-making capability among members. Access to electricity for businesses and commercial establishments developed economic capital. This evolved community capacity helped them to go beyond Bondo-1, and they planned to expand their facility by establishing two more powerhouses and extending their distribution network for broader coverage. Interestingly, the community's capacity is continuously evolving during all these developments, and now they are planning to establish one larger powerhouse to cater for more households and industrial customers.

The community's capacity evolved during all these implementation phases, which led to the subsequent development of more powerhouses or infrastructure and made it resilient against unforeseen events. During the development of the Bondo mini-grid over the last few years, the community encountered many unexpected disruptions and barriers which could be a possible threat to the seamless operation of the mini-grid. For instance, during the establishment of Bondo-2, there was an incident of severe flooding which destroyed the under-construction building of the powerhouse and claimed the lives of two security persons guarding the powerhouse. However, this disastrous event didn't stop them, and the community's evolved resilience enabled them to realise their plan of establishing the Bondo-2 powerhouse again successfully. Besides, there were reported to be a few socio-political disruptions when some people from nearby villages approached Bondo community members and provoked them to demand compensation for placing electricity poles in front of their houses. However, these incidents were handled diligently by energy committee members as community mobilisation happened during project planning, strengthening the social bond and community's resilience to overcome these issues.

2.5 Building Resilient Community Energy Systems

Realising resilient community energy systems provides reliable energy sources to underserved and vulnerable communities and develops local autonomy by reducing their dependency on public infrastructure. For instance, cyclone Ana 2022 in Malawi displaced more than 190,000 people and severely impacted the country's largest hydropower generation facility, which supplies nearly one-third of the country's electricity demand (Department of Disaster Management Affairs, 2022). It took several months to restore the system to its original working condition, and consumers connected to the national grid suffered frequent blackouts and power cuts during that period. However, communities powered through decentralised and community energy-based power generation systems were least affected during that period as they were not dependent on a single electricity utility (national grid) for their power consumption. Also, not all the regions were severely impacted by the cyclone; some regions were marked safe due to their geography. However, people residing in safe regions also had to go through power disruption issues due to instability in the national grid infrastructure. In such events, repairing the damage and restoring the large-scale public infrastructure to its original state may take several months or years. However, restoring a small-scale community energy project is much easier and quicker in case of unforeseen events. Thus, community energy systems induce resilience through their decentralised structure and autonomy.

A resilient community energy system can be developed based on three core aspects—energy system design, community involvement, and governance (Fig. 2.6). These three pillars form a solid foundation to realise a resilient and sustainable CE project and play an essential role in all project implementation and operation phases. Energy system design is a core aspect of strengthening physical or infrastructural capacity to make the energy system resilient against unpredictable events or hazards. Community involvement and governance are the aspects of developing community resilience by supporting the community's inherent capital to cope with unforeseen events or disruptions. These aspects are closely related to previous sections describing the contribution of CES in building

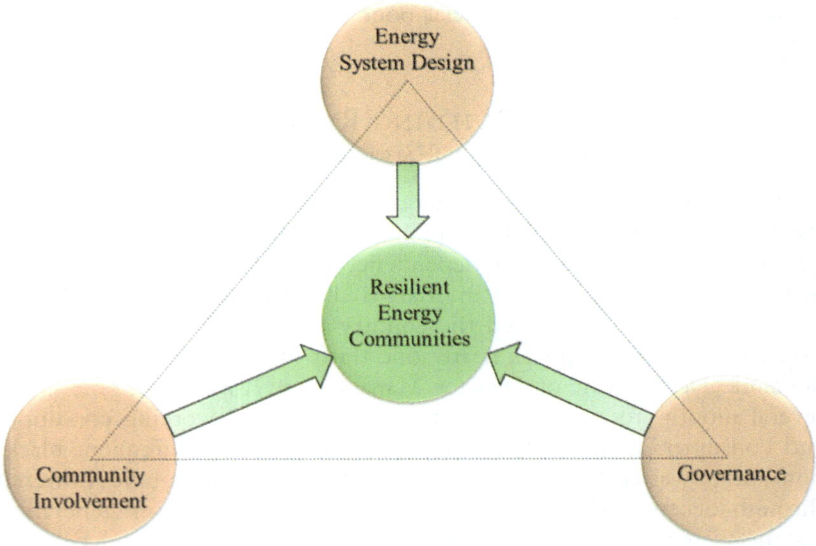

Fig. 2.6 Core components of resilient community energy system (*Source* Author's own)

community resilience and the role of community resilience in defining the long-term sustainability of a project. The three pillars of building a resilient community energy system are described below.

2.5.1 Energy System Design

It mainly concerns the system's technical design, like the type of technology used, equipment class, and other technological standards for developing an energy system. The energy system design plays a significant role in determining the resilience of the overall system. A system installed with robust, efficient, and fault-tolerant devices is believed to be more resilient towards sudden disruptions. It means technological innovations should be introduced in energy systems design for their optimised operation and to make them effective and efficient against unforeseen disruptions.

Moreover, energy systems with redundant critical equipment and diverse energy sources are more resilient than those fed by a single source. For example, a hybrid solar-micro-hydro-based CES is perceived to be more resilient than a sole micro-hydro-based CES, where solar can substitute the shortcomings in the micro-hydro system in case of any disruption. Similarly, a micro-hydro CES with a redundant water channel is more resilient than a single channel system because there will be an alternative path for water to hit the turbine in case of any damage to the channel through landslides or other hazards.

However, adding such redundancies and diversity increases the overall cost of the system. Therefore, a proper assessment of risks needs to be carried out while planning the technical design of the system to identify the critical points where adding redundancy appeared to be worthwhile for the safe and sustainable operation of the system.

2.5.2 Community Involvement

Community participation is the second core aspect of realising a resilient CES. Involving the community in one or all phases of project implementation builds the community's overall resilience towards any unforeseen events. Integration of users and making them active stakeholders in the planning, installation, and operation of an energy system contributes to developing the community capital. Active participation in the community includes people's contributions in terms of money, workforce, or any other assets. It induces a sense of ownership among the people right from the project inception stage, motivating them to give their best to safeguard the system from disruptions.

There exist various business models for realising a CES. Most of the CES involves active community participation during the operation phase and limited participation during planning and installation. Such projects are often funded by external organisations and agencies that provide design and installation grants. Sometimes, when these projects encountered any unforeseen disruption during the operational phase, they failed to cope with those events due to their limited knowledge and ability to handle them. On the other hand, projects involving active community participation in all three phases of project implementation have demonstrated enhanced capacity to cope with any unforeseen disruptions on their own.

2.5.3 *Governance*

Effective governance is one of the significant aspects in the lifecycle of a sustainable CES. Governance includes making policies, laws, and schemes for promoting CES at the grassroots level. More often, it is referred to as an enabling environment facilitating the smooth implementation of new projects and safeguarding the sustainable operation of existing projects. Governance exists at three levels—community, regional, and national. At the elementary level, an energy committee formed to maintain and operate the CES is responsible for community-level governance. The energy committee is the first-hand body in charge of monitoring, maintaining, and using the CES successfully and providing updates to the local or regional governance bodies about the system's health.

Local or regional governance is performed by different institutions such as state or county governments, statutory agencies, or regional development bodies. These institutions are responsible for framing regional policies and schemes supporting CES at the ground level. They provide support by engaging stakeholders like private energy companies, development organisations, and public utilities. Local governing institutions facilitate energy committees to locate funding opportunities for grants, loans, and public investments. Their primary role is to monitor the CES in their jurisdiction and safeguard users' interests by resolving any conflicts among users, developers, or other stakeholders. They act like an intermediary institution for implementing any national-level plan and are also responsible for providing approvals and permissions to project developers.

National-level governing agencies or ministries are responsible for rolling out any country-level policy or scheme supporting the deployment of these projects. They facilitate project developers by introducing national-level plans offering subsidies, loans, tax exemptions, and other regulatory advantages. Additionally, they monitor the status of these projects concerning their national and international agenda of providing sustainable and affordable energy access to all. They even coordinate with international development agencies to secure financial aid or long-term sustainable development credits. To realise a resilient and sustainable CES, there needs to be strong coordination among all the institutions to deliver all these international, national, and regional benefits. Often, improper design and conflict between national and regional policies limit beneficiaries from utilising the intended support. Therefore, effective integration and strong coordination among governing bodies play a major role in realising a resilient and sustainable CES.

These core aspects of realising a resilient community energy system must be incorporated in each stage of CE project implementation, as shown in Fig. 2.6. For instance, when robust and redundant 'energy system design' is included right from the conceptualisation stage, it results in the development of resilient infrastructure which could withstand possible disruptions. Similarly, when aspects of 'community involvement' and 'governance' are considered during the CE project's installation and operation, it enhances the community's social and human capital, eventually contributing to the community's resilience with respect to any unforeseen event or disruption. Thus, incorporating these core aspects makes the energy system resilient and contributes to developing community's resilience.

2.6 Conclusion

The concept of resilience holds a greater significance concerning communities that are poor in terms of infrastructure or physical capital and highly prone to natural and socio-political disruptions due to their geographies. Every year, millions of people are displaced or affected in other ways due to unforeseen natural events or socio-political conflicts. Such incidents destroy the physical infrastructure and severely damage the community's capital. Undoubtedly, it takes so much time and resources to rebuild the damaged infrastructure and several years to restore the community's capacity in the form of social, economic, and human capital. Handling such events through top-down relief and humanitarian approaches provides short-term assistance. However, a long-term solution requires imbibing the aspects of resilience at the ground level within the community. As resilience of a community is the measure of its social, economic, human, and physical capital, strengthening the community's capital is the way to achieve community resilience. A resilient community is one which efficiently utilises its inherent capital during various phases of disaster management.

Community energy systems shares a strong bond with community resilience. Each phase of a CE project implementation contributes to building one or more forms of community's capital. The evolved community's capital is the measure of its resilience against any unforeseen event or disruption. Community energy is such a unique initiative which not only provides clean, reliable, and last-mile energy access to energy-deprived communities but also strengthens the aspects of community resilience, preparing them to cope with unforeseen events. Building resilient community energy systems across underserved and vulnerable communities accomplishes the objectives of SDG-7 (affordable and clean energy) and contributes to SDG-11 (sustainable cities and communities). A resilient community energy system could be realised based on the aspects of 'energy system design', 'community involvement', and 'governance'. Considering these core aspects will deliver a resilient and sustainable energy system and a resilient community. The unique approach of analysing the relationship between CES and community resilience opens the gate for further researchers to investigate possible ways through which community resilience could be strengthened through CES implementation. The novel framework to realise resilient CES based on three core dimensions lays the foundation for future research to develop guidelines for the efficient delivery of such systems.

References

Coaffee, J., & Lee, P. (2016). Urban resilience. *Urban Studies: Planning for Risk, Crisis and Uncertainty, 59*(1), 3–19. https://doi.org/10.1177/00420980211052230

Community and Regional Resilience Institute. (2013). *Definitions of community resilience: An analysis* (Carri Report, pp. 10–11).

Cretney, R. (2014). Resilience for whom? Emerging critical geographies of socio-ecological resilience. *Geography Compass, 8*(9), 627–640. https://doi.org/10.1111/gec3.12154

Dabson, B. (2015). *Planning for a more resilient future: A guide to regional approaches*. NADO (National Association of Development Organization). Research Foundation.

Department of Disaster Management Affairs. (2022). *MALAWI: Tropical Storm Ana, 2*(1), 2–3.

Disasters Emergency Committee. (2019). *2019 cyclone IDAI appeal final report* (pp. 2–3). Retrieved January 9, 2024, from https://www.dec.org.uk/sites/default/files/media/document/2021-10/DEC_Idai%20Report%20final_21_HR_SP_AW.-compressed.pdf

International Federation of Red Cross and Red Crescent Societies. (2004). *World disasters report. Focus on community resilience* (pp. 18–31).

International Federation of Red Cross and Red Crescent Societies. (2014). *World disasters report. Focus on culture and risk* (pp. 93–113). Retrieved January 9, 2024, from https://www.ifrc.org/sites/default/files/WDR-2014.pdf

IPCC. (2022). *Climate change 2022: Impacts, adaptation and vulnerability. Working group II contribution to the IPCC sixth assessment report* (pp. 35–36). https://doi.org/10.1017/9781009325844

Kuhlicke, C., Madruga de Brito, M., Bartkowski, B., Botzen, W., Doğulu, C., Han, S., Hudson, P., Nuray Karanci, A., Klassert, C. J., Otto, D., Scolobig, A., Moreno Soares, T., & Rufat, S. (2023). Spinning in circles? A systematic review on the role of theory in social vulnerability, resilience and adaptation research. *Global Environmental Change, 80*, 102672 https://doi.org/10.1016/j.gloenvcha.2023.102672

Millennium Ecosystem Assessment. (2005). *Ecosystems and human wellbeing: Synthesis* (pp. 4–6).

Mayunga, J., & Peacock, G. W. (2010, December). The development of a community disaster resilience framework and index. In *Advancing the resilience of coastal localities: Developing, implementing and sustaining the use of coastal resilience indicators: A final report* (pp. 2–57). https://doi.org/10.13140/RG.2.2.23402.75201

National Geographic Society. (2023). *Mar 11, 2011 CE: Tohoku Earthquake and Tsunami* (pp. 1–2). Retrieved January 9, 2024, from https://education.nationalgeographic.org/resource/tohoku-earthquake-and-tsunami/

Norris, F. H., Stevens, S. P., Pfefferbaum, B., Wyche, K. F., & Pfefferbaum, R. L. (2008). Community resilience as a metaphor, theory, set of capacities, and strategy for disaster readiness. *American Journal of Community Psychology, 41*(1–2), 127–150. https://doi.org/10.1007/s10464-007-9156-6

Oloruntoba, R., Sridharan, R., & Davison, G. (2018). A proposed framework of key activities and processes in the preparedness and recovery phases of disaster management. *Disasters, 42*(3), 541–570. https://doi.org/10.1111/disa.12268

Ortiz, W., Dienst, C., & Terrapon-Pfaff, J. (2012). Introducing modern energy services into developing countries: The role of local community socio-economic structures. *Sustainability, 4*, 341–358. https://doi.org/10.3390/su4030341

Peacock, W. G., Brody, S. D., Seitz, W. A., Merrell, W. J., Vedlitz, A., Zahran, S., Harriss, R. C., & Stickney, R. (2010). *Advancing resilience of coastal localities: Developing, implementing, and sustaining the use of coastal resilience indicators: A final report*. Hazard Reduction and Recovery Center, Texas A&M University.

Tierney, K. (2015). Resilience and the neoliberal project: Discourses, critiques, practices—And Katrina. *American Behavioral Scientist, 59*(10), 1327–1342. https://doi.org/10.1177/0002764215591187

To, L. S., & Subedi, N. (2019). Towards community energy resilience. In *Energy access and forced migration* (pp. 81–91). https://doi.org/10.4324/9781351006941-8

White, R. K., Edwards, W. C., Farrar, A., & Plodinec, M. J. (2014). A practical approach to building resilience in America's communities. *American Behavioral Scientist, 59*(2), 200–219. https://doi.org/10.1177/0002764214550296

World Health Organization. (2023). *Tsunamis* (pp. 1–2). Retrieved January 9, 2024, from https://www.who.int/health-topics/tsunamis#tab=tab_1

CHAPTER 3

Where Is the Community in Community Energy?

Hita Unnikrishnan

Abstract One of the most enduring debates in relation to community energy is the significance of the term community. Community energy refers to the active role that people may play in managing and governing energy resources and technologies, but the community is not an homogeneous unit that can be easily identified and defined. Instead, this chapter argues for recognising the inherent diversity of the communities that participate in community energy and how that diversity enhances their potential, but also causes challenges. The chapter explores different theorisations of communities—communities of interest and communities of practice—as well as the different imaginaries associated with the notion of community. This analysis reinforces a dynamic perspective on ideas of community and their mobilisation within particular political contexts.

Keywords Community of practice · Community of interest · Diversity · Identity · Partnerships

H. Unnikrishnan (✉)
Urban Institute, University of Sheffield, Sheffield, UK
e-mail: h.unnikrishnan@sheffield.ac.uk

V. Castán Broto (ed.), *Community Energy and Sustainable Energy Transitions*, https://doi.org/10.1007/978-3-031-57938-7_3

3.1 Introduction

Community energy is an alternative for decarbonisation, energy access and resilience. There are two distinct ways within which to approach the idea of community energy. On one hand, the notion of energy communities refers to organised ways in which groups of people provide themselves with energy. This approach is becoming more popular particularly in the context of energy transitions occurring within the European Union and forms the backbone governing small renewable energy systems in the region. On the other hand, the notion of community energy systems focuses on the organisation of communities to facilitate the generation of electricity. In both cases, the idea of community energy maps a heterogeneous array of decentralised models to manage energy provision.

Both these approaches engage deeply with the complex term 'community'. The notion of 'community' has been increasingly used to design intervention programmes across diverse contexts—for example architectural projects, planning guidance, or development planning initiatives. However, 'community' as defined in these initiatives implicitly or explicitly black boxes the notion into a description of homogeneous individuals driven by a common shared interest in the landscape of intervention (see Rigon & Castán Broto, 2021, among many others). 'Community' is often linked to unrealistic portraits of idyllic groups of people living in harmony with their landscapes (Brint, 2001). Behind those idealistic fantasies, there is a complex landscape consisting of diverse groups of people living in proximity to each other—proximity that may be physical or ideological (Harvey, 1997). Communities are thus shaped by internal power relations, by complex institutional arrangements, as well as by their own individual interpersonal interactions with and between each other.

These inherent complexities to those communities are magnified when we think of community energy because it represents an interaction between complex communities and the equally complex socio-technical systems that support them. Community energy systems provide alternative models to the prevalent systems of privatised off-grid energy provision, in which infrastructure systems are managed by businesses. At the same time, communities may not have the consolidated structure of a private company. How then do they enable the management and governance of energy projects? To address these questions, it becomes important to unpack the notion of 'community' within community energy systems and what that means for how community energy projects are operationalised on ground.

This chapter reviews the notion of "community" as it is prevalent in various strands of scholarship and how criteria for inclusion and exclusion in these "communities" vary in subtle or distinctive ways. This helps explore the question of where the 'community' in community energy systems is and how communities are envisaged and conceptualised in the context of community energy projects. Subsequent sections offer an outline of the critical theory on community as well as different conceptualisations of the notion of community. The chapter focuses on three ideations of the word community—as a partnership between people with different objectives, as a nationalist-linked institution, and as a place-bounded group of people. The analysis reveals some of the pitfalls that exist in community development—including the risks surrounding elite capture, new forms of oppression, post-political decision making, and conflict. Nevertheless, the concept of 'community' has been mobilised in different discourses around community energy, including within academic literature and policy reports. The chapter goes on to demonstrate how community is an important gel that enables community-centric projects. At the same time, communities are dynamic. The term community needs to be interpreted in ways that seek to foster inclusion and dialogue between diverse groups of individuals rather than promoting the currently dominant homogeneous, exalted perspective of shared common ideals.

3.2 Understanding Community

A literature review of how the term community is deployed across scientific literature shows that the word 'community' is often used in both value-laden and unclear ways. This is consistent with previous literature on the topic—see for instance Head (2007) who describes the term as 'notoriously vague and value-laden' or Bryson and Mowbray (1981) who criticised the use of the term as a "spray-on solution" to address diverse societal and economic issues. At the same time, newer ideas of communities are constantly being mobilised, politicised, and perpetuated by an equally diverse range of actors—for instance emergent notions of energy communities, policy communities, or heritage communities (Alatalu, 2021; Caramizaru & Uihlein, 2020; Miller & Demir, 2007). Brosius and Hitchner (2000) call for greater critical reflection on the imaginaries of cultures, communities, and practices as being homogenous and eternally enduring and the positive and problematic impacts of such imaginaries on policy and practice.

The most dominant ways of using the term 'community' in contemporary literature include communities of interest and practice or action. The term 'community' further carries with it associations of multiple imaginaries—three of the most dominant ways in which the term is reimagined are—(a) to enforce an imagined oneness of nation, (b) to construct a sense of homogeneity based on geographical proximity, and (c) to envision partnerships of individuals coming together with a shared and common objective. Each of these forms of communities is not mutually exclusive and can often intersect within and across each other.

3.2.1 Communities of Interest

Scholarly literature on communities of interest has emerged from diverse disciplines—for instance from computer science and design (Fischer, 2001) to within critical social sciences and anthropology. Described as 'communities of communities' (Brown & Duguid, 2001), communities of interest are characterised by their shared interest in framing and resolving a particular problem (Harrington et al., 2008), are considered more transient or temporary in comparison to other forms of communities (Fischer, 2001). The assumption here is that while there is commonality in the objectives of the community, understanding the nature of the problem and developing a potential solution together is an incremental process, dependent upon effective communication between and across community members enabling the sharing of diverse skills and perspectives (Fischer, 2001). A core requirement for such communities to function effectively is the establishment of adequate boundaries through the establishment of boundary objects (Star, 1989) that have meaning across the various individual knowledge systems represented by a community of interest. Kelly (1996) identifies "transcendent" communities of interest characterised by groups located in multiple spaces that are linked beyond their physical geographies.

In more practical terms, communities of interest are often defined in terms of their common economic considerations or their belongingness to a shared industry (Rosenfeld & Duchin, 2022), thereby invoking calls for their inclusion within urban policy, governance, and practice—for instance to draw boundaries for districts within a jurisdictional area. Communities of interest have been posited to have multiple benefits—for instance their potential to represent marginalised views and present an alternative to dominant neoliberal discourses around resource governance and practice (Harrington et al., 2008; Mercer & Marden, 2006).

3.2.2 Communities of Practice

Communities of practice have been defined variously by different schools of thought, yet they all seem to converge on the idea of communities that demonstrate a sustained level of participation, engagement, and collaboration around a particular domain that revolves around a common theme with marked boundaries—for instance architecture and urban planning (Fischer, 2001). Wenger (1998) defines a community of practice as consisting of groups of individuals who interact with both each other as well as the world, towards pursuing an enterprise co-created and perpetuated by the group. A core assumption here is that these communities operate around a single knowledge system towards which people progress with greater experience built over time. Communities of practice are particularly distinguished by the presence of a shared practice between participants and the fact that they are based on rational interest as opposed to the forging of emotional ties between them (Wenger, 1998).

Studies have suggested however that practice can mean different things to different people and these subtle differences in the way the term is understood by diverse actor groups can often affect both participatory processes as well as the communities that emerge from them (Buchy & Race, 2001). Nelson and Wright (1995) distinguish between two different kinds of participatory processes based upon the purpose for which it is adopted. These include instrumental participation—where the tool is used to meet a specific goal; and transformation—where participation is used as a tool to deliver on social change objectives. Communities of practice therefore can differ even within themselves based on their intrinsic motivations and the objectives contained within their activities.

3.2.3 Imaginaries Associated with "Community"

The term community is associated with some distinctive imaginaries. Three of the most dominant ones are presented below:

3.2.3.1 Communities and Identity

Communities of identity emerge from socio-cultural ties that go beyond geographical proximity, but are formed through specific experiences, place associations, or practice giving rise to communal belonging and identification (Duane, 1997). One example of such communal identities is nationalism which gives rise to a form of mutual oneness, that can very well be defined as a community. Nationalism needs to be distinguished here from the actual emergence and formation of a nation state as an intangible consciousness permeating the collective identities of people who identify with it. Anderson (1983) argues that "nationality/nationness as well as nationalism are cultural artefacts of a particular kind", one where the nation becomes an "imagined political economy"—both continuous through time as well as bounded by space (Stephens, 2013). These communities are "imagined" because members of even the smallest sovereign unit are unlikely to know or interact with most of their 'community'. However, a sense of solidarity is constructed between these individuals based on (largely print media-propagated) myths of shared and common histories, and therefore a shared and common future (Sarsembayev, 1999). This of course raises further questions about the scale of this imagination, the modes in which it occurs, and whether any group of individuals possessing the right characteristics may be considered 'imagined' communities (Yael, 1995).

Imagined communities here are not defined by whether they are a group of individuals genuinely coming together for a shared common purpose, but rather by their imagination driving them towards constructed solidarity. Yael (1995) further distinguishes imagined communities as constructing mutual solidarities as based on beliefs of common ancestry, fraternity, genealogy, distinctiveness, and exclusivity (Yael, 1995). This notion of constructed communities is not fixed, but inclusion and exclusion within them is dynamic and dependent upon changing geopolitical conditions including immigration, teleconnections, recognition of gendered identities, or even mixed-race unions and global conspiracies (Malešević, 2020; Yael, 1995). Gellner takes a slightly different approach in arguing that nationalism is a product of social change, modernisation, and a growing awareness of class especially among the worker classes and intellectuals leading to their alienation from tradition and bringing them together to create a new or alternative sense of cultural homogeneity (Gellner, 1983). Smith goes on to state that nationalism as a force binding individuals cannot be understood without grounding the phenomenon in the idea of a 'national identity' which is collectively constructed (Smith, 1991).

In this context, studies that explore the emergence of nations and the emergence of nationalistic tendencies have argued that the romantic image of homogeneous, natural, and historical continuous communities as posited by nationalists are in fact mere mirages that do not reflect realities on ground (Yael, 1995). Other studies question the validity of the notion of imagined communities particularly in the context of countries such as India and those in Africa or Latin America that have colonial legacies or otherwise diverging paths towards the development of a nation state. Such studies posit that nationalism and the idea of a 'modular national society' are largely Western constructs—one that emerged through processes of colonialism that suppressed the institutions and political spirit of the colonised and in this sense, 'imagined communities' are also examples of 'colonised imaginations' (see for instance Chatterjee, 1991; Go & Watson, 2019; Itzigsohn & vom Hau, 2006). Recent studies (for example Cowan, 2021) have argued that communities (such as nations) do not exist out of nothing, but rather they emerge through human capacities to interact, communicate, and imagine with one another, and the emergence and subsequent roles played by new forms of media (such as the social media and indeed the internet) in creating new communities is critical.

3.2.3.2 *Communities and Place*

Another prominent school of thought links communities to place and geography. The earliest strand of literature examining the notion of community through a place-based lens is from the work of Tönnies and Loomis (1957) who coined the term '*gemeinschaft*' for closely knit and intimate social relations bounded in a well-defined geographical space. This work together with that of Durkheim (1964) established the notion that communities are normatively a good thing, and that societal breakdown occurs because of the breakdown of these units. Lee and Newby (1983) typified communities into locality-based communities (where the commonality is limited to shared physical spaces), local social systems (based on the interactions between individuals), and communion (where individuals have a shared sense of identity). Subsequent literature has highlighted the importance of focusing the communal identity around familiar geographical features, cultural landmarks, or familial connections especially in the context of their roles in facilitating interactions between individuals (Hargreaves, 2004; Nash & Christie, 2003; Robertson et al., 2008).

Such discourses create an imagination of the community as being bounded by their proximity to and commonality in interactions with specific geographies—for example around sites of environmental contamination (Edelstein & Edelman, 1988), or around landscapes of ruination, decline, or deprivation (Mah, 2012). It brings to light tensions around whose sense of place are we engaging with and what claims to place are included or excluded in these narratives (Cameron, 2003; Harrington et al., 2008). Several strands of literature we have examined do not explicitly define what they mean by the term community. The word may be used to refer to individuals who do not participate in or are excluded from development, nationalistic or place-based narratives (for instance see Skerrat & Steiner, 2013). This perspective challenges the normatively held notion that one of the prerequisites to identify functioning communities is a healthy degree of participation (Gilchrist, 2009). It further implies that the formation of a community does not require action, but that individuals can be cohesively bounded through their inaction (communities of inaction) or conversely that action of some form (such as community-led efforts or developmental initiatives) has the potential to form new 'communities of exclusion'. Stratford et al. (2003) further note that local places can fall victim to barriers restraining the objectives of community development—for instance elitism or parochialism, while at the same time paying insufficient attention to complexities such as intersecting community memberships or vulnerabilities (Smith, 2004). They can act as boundary making processes that both separate as well as divide people and places based on geographic nearness (Blomquist & Schlager, 2005; Harrington et al., 2008).

3.2.3.3 Communities as Partnerships with a Shared Objective

A third strand of literature where communities take centre stage is the discourse on collective action, management of the commons, and co-production. Communities here are defined as having a few shared characteristics—a degree of stability through regular or semi-regular interactions between members on several fronts, relationships that are largely not state-mediated, and who share beliefs and preferences that go beyond the problem requiring collective action (Flora, 1998; Ostrom, 1990; Taylor & Singleton, 1993). The distinction here, however, is that inclusion in such communities does not necessarily involve strong feelings (including altruistic ones) among members of the group (Flora, 1998). Such communities are described as being a sort of social infrastructure that differ along different geographies as well as in their acceptance of different ideas or people (Flora, 1998)—differences that affect their ability to make decisions that eventually benefit developmental goals (Wilkinson, 1991). Conflict management strategies are therefore considered essential prerequisites of interactions within such communities (Ostrom, 1990). Further, literature emerging from here explicitly highlights the challenges and trade-offs between communities and the state—for example drawing attention to the limits of community capacities and ability for decision making (Singleton, 2000). In so doing it points to the importance of institutional arrangements that strike a balance between state and community objectives (Cox et al., 2010; Singleton, 2000; Unnikrishnan et al., 2023).

This school of thought centres ideas of differences between communities—there is recognition of the fact that communities can be composed of different sub-groups, each with equally divergent capacities to act (Shucksmith, 2000). They are further defined through heterogeneities in their perceptions, resources, influence, as well as interests in engaging with the collective action problem (Crona & Bodin, 2006). While the recognition of this diversity is clear within this literature, there is debate on whether this kind of heterogeneity is beneficial in achieving the goals of developmental or community-led initiatives. For example, the design principles for collective action emerging from within the Bloomington school of institutional thought (Cox et al., 2010; Ostrom, 1990) imply that heterogeneities within communities might indeed pose barriers to collective action. Conversely, Newman and Dale (2005) argue that while heterogeneity might seem counterproductive in the initial stages of a project, it might indeed be a key towards increasing the adaptive capacity and resilience of these groups to external shocks. Recognition of intra-community heterogeneities is however limited within discussions emerging from this standpoint.

Literature emerging here underscores the notion of temporality in the process of community formation—studies have demonstrated processes of community formation are non-linear and may depend largely on everyday interactions between individuals in contrast to the prescriptions posited by development practitioners and policy analysts (Khotari, 2001).

3.3 Challenges and Limitations on Building Communities

The notion of community brings with it a particular contradiction—how do we bring different people together across barriers posed by differences between individuals in ways that address systemic inequalities (Rosenfeld & Duchin, 2022)? In the words of Staeheli (2008), communities can risk the "construction of sameness, rather than a recognition of what is common"—it can totalise instead of being constituted through sharing the dynamicities of communities—the notion that communities can be lost from its original environment (village or local neighbourhood), and regained through shared space, ideologies, and values (Driskell & Lyon, 2002). Rather communities progress from one state of being into another (Driskell & Lyon, 2002). Why does community matter?

The question of how people come together in a place and why has many answers, yet these explanations are not always explicitly or consciously articulated (Rosenfeld & Duchin, 2022). Further conflicts can arise at the intersection of different kinds of communities causing what has been termed community risk (Duane, 1997; Taarup-Esbensen, 2019). For instance, Taarup-Esbensen, in documenting the community relations negotiated by mining communities, describes how communities of place (described as people living near the mine and who are directly affected by it) intersect with communities of interest (defined as people outside of the place who either support or oppose the activity) in either cooperative or conflictual ways thus influencing broader narratives about the activity in general. In the next section we further problematise these notions of community by taking each of the types of "community" we described in the previous section and elucidating the politics of inclusion in those contexts.

In Sects. 3.1 and 3.2 we presented how the idea of "community" varies according to the purpose and motivations for which they have been formed. We now discuss the politics of inclusion within "communities" using the examples presented earlier. For each kind of community, we discuss the conditions for inclusion and conversely the boundaries on inclusion placed by those conceptualisations of the term community.

As may be seen in Table 3.1, each framing of the term community carries with it implicit assumptions about what constitutes belongingness as well as the boundaries associated with belonging. Paying attention to these distinctions can help shed light on groups that get excluded within each framing of the notion, thereby allowing for community-led initiatives to incorporate better notions of equity and justice in their activities. For instance, it may be seen that each of these framings of community serves to homogenise individuals in different ways—whether they be through imagining a shared common identity or establishing partnerships for a fixed and defined purpose. Indeed, literature that focuses on collective action goes a step further in stating that heterogeneity may in fact hinder successful collectives in forming (Cox et al., 2010; Ostrom, 2005) and operating efficiently. In so doing, they explicitly tend to exclude those individuals who do not conform to these shared ideals and imaginaries creating deep implications for society at multiple scales (for instance creating politically divided societies, or unequal benefits reaped by different actors through community-led initiatives like energy).

3.4 "Community" as Operationalised in Policy Reports

While scholarly literature has posited several framings of the term "community", the term has become an important buzzword in many policy fora across diverse sectors. Communities are seen as lynchpins towards successful governance—particularly with regard to their abilities to engage with decentralised infrastructure provisioning or its monitoring. Communities are posited as being critical in developing a region's adaptive capacity to shocks and vulnerabilities, thereby increasing its social and ecological resilience. At the same time, each of these policies varies in terms of what constitutes a community—in many cases the community itself remains undefined. This is particularly true within policy framings of community energy—a concept that has gained considerable traction within the Western world, and whose practices have spilled over into major international development agendas (such as the Overseas Development Aid funds).

3.4.1 Policy Landscapes of Community Energy

A comparative analysis of key reports on community energy helps situate the notion of community within community energy and the key challenges they raise (Table 3.2). The reports include a selection of policy documents that attempt to support the role of community energy in the energy transition.

Table 3.1 The politics of community

Type/imagination of community	*Conditions for inclusion*	*Benefits of inclusion criteria*	*Associated exclusions*	*Implications*
Communities of interest	Shared interest in framing and resolving a problem	Brings together people with a shared and common approach to a problem	Differences in motivation, economic or disciplinary boundaries	Creation of a sense of homogeneity with respect to the problem at hand, while exclusion of people who do not conform to the belief occurs passively
Communities of practice	Participation, Engagement, Collaboration	Brings together diverse people who work together actively towards developing a solution to an identified problem and potentially reap benefits through that activity	People with interest but who do not have the motivation or ability to participate actively	While diversity in perspectives is embraced, this framing excludes people who may not be able to participate or collaborate as actively as the project requires, thus keeping them from also obtaining benefits that emerge from the partnership. This can have deep implications for systems like those of community energy, particularly in energy deprived regions

Type/imagination of community	*Conditions for inclusion*	*Benefits of inclusion criteria*	*Associated exclusions*	*Implications*
Communities and identity	Imagined belonging to shared identity based on presuppositions of what constitutes that identity	Creates a sense of belongingness to a larger project—such as nationalism or communal identities	People who do not conform to the presupposed identifying markers of that identity	Can create deeply divisive structures, fosters the notion of us versus them as is seen in politically divisive societies of the present day
Communities and place	Proximity to or deep-rooted ties to a particular geography	Brings together diverse people who identify as belonging to a sharply defined geographical space	May be exclusionary to recent migrants into a landscape or transient individuals (by virtue of them not possessing the adequate markers of belongingness)	Can exclude certain sections of the society—particularly those who are the most vulnerable
Communities as partnerships	Shared sense of purpose reinforced through regular interactions without strong emotional ties to the partnership	Creates a partnership towards a defined purpose without the sense of kinship not extending beyond that defined purpose	May exclude those individuals whose approach to the purpose of the partnership does not conform to the dominant world view	Excludes difference, creates a sense of homogeneity by virtue of this framing

Table 3.2 Policy framings of "community" in community energy

Report	*Key point of the report*	*Mentions of community*	*Definition of community*	*Identification of challenges dealing with community*
Environmental Protection Agency, Ireland, Responding to the Energy Transition in Ireland: The Experience and Capacity of Communities report, 2020	Explores the Irish experience of community energy, support, and development of community capacity for energy transitions and the role of intermediary groups in supporting community led responses to the energy transition	625	Defines community as "a collective actor with agency that can interact with others". Draws attention to other meanings of the term community. Proposes a shift away from what a community is to what its role is in the energy transition. In describing results from their case studies, acknowledges the diversity of perceptions surrounding term from beneficiaries and stakeholders of community energy projects	Yes. Mentoring relating to community development; need for outside and ongoing help in finance, advice, education, and guidance. Need for intermediary groups to act as go-betweens. Trade-off between the ambition of energy citizenship and the struggles faced by communities. Need for clarity on financial and infrastructural barriers to community energy projects before talking up community ownership of energy. Community energy does not automatically guarantee acceptance of the project by local people in the area

Report	*Key point of the report*	*Mentions of community*	*Definition of community*	*Identification of challenges dealing with community*
UK Community Energy, State of the sector report, 2022	Describes the current scenario of community energy in the UK and identifies challenges associated with delivering the promise of community energy—from policy-level barriers to the impact of COVID-19 on the sector	240	Does not define community or community energy in any way within the policy brief. Various terms such as community organisation, community-owned, community shareholder appear through the document with little clarifying explanation as to what distinguishes these actors from one another	Yes—some of the challenges identified include (a) lack of organisational capacity within community energy projects, prevalence of voluntarily undertaken labour, and the presence of multiple intersecting roles and interests of community groups of which energy is just one part
Department of Energy and Climate Change (Now defunct Department of the UK Government) Community Energy Strategy, 2014	Represents the first community energy strategy for the United Kingdom and places communities at the centre of the energy transition	946	Does not define what a community is though it acknowledges that communities can be of different types—a group of people setting up a solar or wind turbine, a local authority leading a collective purchasing scheme, an energy advise session at a local community centre or other schemes. Implicitly the document seems to tend towards framing a community as a partnership with a shared vision	Yes. Prevalence of voluntary labour, and therefore maintaining enthusiasm for the project among wider members of the public was identified as a major challenge

(continued)

Table 3.2 (continued)

Report	*Key point of the report*	*Mentions of community*	*Definition of community*	*Identification of challenges dealing with community*
United Nations Leveraging Energy Access for Advancing Sustainable Energy Goals, 2021	Forms a policy brief demonstrating the interlinkages between SDG 7 (on energy access) and all the other SDGs, as well as includes regional perspectives demonstrating these linkages from each part of the globe	44	Does not define what constitutes either community or community energy, though points to different actors within the community energy landscape	No
Practical Action policy paper Poor People's Energy Outlook, 2016	Policy brief that puts people at the centre of energy planning through community-driven access plans. Reports pilot projects conducted in several energy-poor nations and their implications for national energy plans	116	Does not define what a community is composed of. Different kinds of communities are identified however such as the global development community and community groups	No

Report	*Key point of the report*	*Mentions of community*	*Definition of community*	*Identification of challenges dealing with community*
Energicentrum (Sweden) Developing community energy initiatives: a literature review (Liljenfeldt & Soares, 2020)	Represents a pre-study for a pilot region in Sweden for the energy transition—Gotland. The study was undertaken by Energicentrum—the entity serving energy infrastructure in Gotland	84	The review notes how the concept of "community" is integral to the notion of community energy and states that within boundaries of these projects, communities can refer to communities of place or interest and that each of these distinctions brings with it notions of process, outcomes, and benefit sharing. The report goes on to describe how ideas of community have shifted in the discourse around Sweden's energy transition—from local communities to more networked ones through dissemination and outreach activities associated with community energy projects	Yes—challenges of establishing political motivations, trust, and reciprocity among community members, the need for project champions, and the need for intermediaries

(continued)

Table 3.2 (continued)

Report	*Key point of the report*	*Mentions of community*	*Definition of community*	*Identification of challenges dealing with community*
International Renewable Energy Agency, Stimulating investment in community energy: broadening the ownership of renewables, 2020	Provides a showcase of best practices and financing mechanisms towards accelerating community energy development and its benefits	416	Community in this report is synonymous with the homogenous category of citizens—therefore community energy refers to citizen-driven renewable energy projects where citizens take on active roles in production, distribution, transmission, and consumption of energy. Mentions other kinds of communities such as financial communities	Similar challenges identified as in previous reports in this review—voluntary nature of participation, need for champions, and motivation, in addition to the financial barriers to community energy which forms the basis of this report

The analysis in Table 3.2 shows that the idea of community is central to policy briefs developed around community energy and energy transitions in general. Communities are seen as central towards advancing the energy transition in decentralised ways that build capacity while at the same time drawing multiple benefits from such projects. At the same time, despite the multiple mentions of the term in each policy brief we have reviewed, there is a surprising lack of clarity on what "community" within community energy systems means. Communities are very rarely defined for the purposes of each report, and often represent a buzzword that drives the report's agenda forward. Another interesting observation we made through this review was that while communities are expected to self-organise (with or without the presence of intermediary agencies) in the context of community energy systems, little attention is paid to how those communities self-organise—and what their criteria for inclusion are. In other words, there is little scrutiny of internal heterogeneities within groups of individuals coming together for a project and community thus becomes a monolithic and homogeneous concept, externally and ironically imposed by entities to describe a group of individuals participating in the community energy landscape. This framing of the community masks any intersectional vulnerabilities or strengths possessed by the group as well as the loci of power as centred around these groups. However, as several strands of literature have shown, it is important to centre these differences within the community to achieve the goals of justice and equity within energy transitions and to minimise exclusionary practices or the perpetuation of systemic forms of oppression within such initiatives. We therefore recommend that policy begins to pay greater attention to the dynamics of "community" within community energy systems as they have important implications for the delivery of regional and national energy milestones across the globe.

3.5 Lessons for Community Energy

Lazdins et al. (2021) state that energy communities collectivise diverse energy-related actions (like generation, distribution, storage, and pricing) through the agency of local members or stakeholders, thus decentralising the processes of decision making in energy landscapes. In recounting the barriers posed to community energy systems the paper goes on to argue that energy communities face a particular challenge through socio-cultural, political, and organisational means—especially a lack of experience in setting up communities. The implication here is that communities can be artificially created around energy needs with or without external intervention—a notion that stands in contrast to realities on ground that have been enumerated earlier. In developing a multi-player game for decision making within energy communities, Brakovska et al. (2023) define them as "associations voluntarily established by citizens with a common interest in implementing energy efficiency measures and introducing renewable energy sources to reduce their consumption, and energy costs, and increase self-sufficiency".

In a review of business models associated with energy communities, Iazzolino et al. (2022) note that the meaning of the term 'energy community' is variable, involving diverse configurations, actors, technologies, and complexities of infrastructural management. However, they note that energy communities are associated with the joint involvement of citizen groups, entrepreneurs, bureaucrats, and community organisations towards investments they make in the production, distribution, and sale of renewable energy (Iazzolino et al., 2022). The paper further identifies specific challenges associated with community involvement in such projects—their awareness of their own roles in the energy landscape, the propensity of community members to act as passive consumers, their willingness to pay extra for locally produced renewable energy, a lack of access to energy, trade-offs between energy efficiency and the ideals of prestigious and comfortable lifestyles, cost–benefit issues between investors and beneficiaries, energy poverty, and the high capital investment required to finance these initiatives.

In essence energy communities are described as a particular kind of community of interest that has emerged around shared concerns around energy access and appropriation, while challenging dominant discourses of centralised energy production (Helm & Mier, 2019). Literature focuses on the formation of energy communities as being a key movement in driving energy transitions especially in Europe and Australia at least since the 1970s (Iazzolino et al., 2022). Communities are externally defined entities that coalesce around the shared objective of delivering the promise of energy transitions. However, what communities consist of remains vague.

3.6 Conclusion

In this chapter, we have provided a dynamic perspective on the idea of community as operationalised within the community energy sector. To do so, we first expanded on what community has come to signify across diverse strands of literature and the kinds of communities so described. We reflected on the implicit criteria for belonging to these different kinds of communities before examining how the concept of community is operationalised within the burgeoning discourse around community energy systems. As our analysis shows, each conceptualisation of the term community carries with it implicit assumptions about who is included and conversely who gets excluded within them. Our review further noted that while the term has been employed extensively across policy and academic discourses, there is very little clarity as to what the term means on ground in such projects. There is very little engagement with what constitutes a community, what the criteria for belonging to an energy community are, and conversely who do community energy projects thus exclude. Community energy systems in practice and across policy are largely conceptualised as communities of interest operating around a business model of energy infrastructure which can exclude those individuals who either cannot participate in the project or who do not have the means to invest in it. Given that community energy projects are seen as integral to achieving universal energy access especially in energy poor regions of the globe, such framings of community energy projects can in practice implicitly reinforce systemic forms of exclusion. While in practice, communities are expected to self-organise towards developing and consuming community energy infrastructure, their boundaries are often imposed by external entities (such as intermediary organisations, the state, academic researchers, or a funding mandate) in ways that homogenise diverse individuals into either a community of place or interest. These boundaries are reflective largely of the actor/s imposing the boundary but in practice can mask differences in energy access emerging from differential power equations, systemic practices of oppression resulting from class, caste, or gendered dynamics existing within and across such groups of individuals (see, for instance, Kumar & Aiken, 2020). Further, the contrasting perspectives of policy reports and the specific cases suggest that there is a more utopian perspective on community energy in policy, perhaps because the difficulties of applying it in practice are not always evident. Yet, we argue that greater attention needs to be paid to these dynamics of practice as they have important implications towards delivering sustainable and equitable energy transitions across the globe.

References

Alatalu, R. (2021). Dignity of the heritage and the heritage communities. *Journal of Cultural Heritage Management and Sustainable Development, 11*(2), 146–154.

Anderson, B. (1983). *Imagined communities: Reflections on the origin and spread of nationalism*. Verso Press.

Blomquist, W., & Schlager, E. (2005). Political pitfalls of integrated watershed management. *Society and Natural Resources, 18*(2), 101–117.

Brakovska, V., Vanaga, R., Bohvalovs, G., Fila, L. & Blumberga, A. (2023). Multiplayer game for decision-making in energy communities. *International Journal of Sustainable Energy Planning and Management, 38*. https://doi.org/10.54337/ijsepm.7549

Brint, S. (2001). *Gemeinschaft* revisited: A critique and reconstruction of the community concept. *Sociological Theory, 19*(1), 1–23.

Brosius, J., & Hitchner, S. L. (2000). Cultural diversity and conservation. *International Social Science Journal, 61*(199), 141–168.

Brown, J. S., & Duguid, P. (2001). Organisational learning and communities of practice: Towards a unified view of working, learning, and innovation. *Organization Science, 2*(1), 40–57.

Bryson, L., & Mowbray, M. (1981). 'Community': The Spray-on Solution. *Australian Journal of Social Issues, 16*(4), 255–267.

Buchy, M., & Race, D. (2001). The twists and turns of community participation in natural resource management in Australia: What is missing? *Journal of Environmental Planning and Management, 44*(3), 293–308.

Cameron, J. I. (2003). Educating for place responsiveness: An Australian perspective on ethical practice. *Ethics, Place and Environment, 6*(2), 99–115.

Caramizaru, A., & Uihlein, A. (2020). *Energy communities: An overview of energy and social innovation* (Vol. 30083). Publications Office of the European Union.

Chatterjee, P. (1991). Whose imagined communities? *Millennium: Journal of International Studies, 20*(3), 521–525

Community Energy (England, Scotland and Wales). (2022). *State of the sector report*. Retrieved December 11, 2023, from https://communityenergyengland.org/files/document/626/1655376945_CommunityEnergyStateoftheSectorUKReport2022.pdf

Cowan, J. (2021). The constitutive rhetoric of late nationalism: Imagined communities after the digital revolution. *Rhetoric Review, 40*(2), 183–197.

Cox, M., Arnold, G., & Tomás, S. V. (2010). A review of design principles for community-based natural resource management. *Ecology and Society, 15*(4). https://doi.org/10.5751/ES-03704-150438

Crona, B., & Bodin, Ö. (2006). What you know is who you know? Communication patterns among resource users as a prerequisite for co-management. *Ecology and Society, 11*(2). https://doi.org/10.5751/ES-01793-110207

DECC (Department of Energy and Climate Change). (2014). *Community energy strategy: Full report*. Retrieved December 11, 2023, from https://assets.publishing.service.gov.uk/government/uploads/system/uploads/attachment_data/file/275163/20140126Community_Energy_Strategy.pdf

Driskell, R. B., & Lyon, L. (2002). Are virtual communities true communities? Examining the environments and elements of community. *City and Community, 1*(4), 373–390.

Duane, T. P. (1997). Community participation in ecosystem management. *Ecology LQ, 24*, 771.

Durkheim, E. (1964). *The rules of the sociological method*. The Free Press.

Edelstein, M. R., & Edelman, M. (1988). *Contaminated communities: The social and psychological impacts of residential toxic exposure*. Avalon Publishing.

EPA (Environmental Protection Agency). (2020). *Responding to the energy transition in Ireland: The experience and capacity of communities*. Research Report 337. Retrieved December 11, 2023, from https://www.epa.ie/publications/research/climate-change/Research_Report_337.pdf

Fischer, G. (2001). *Communities of interest: Learning through the interaction of multiple knowledge systems*. Proceedings of the 24th IRIS Conference, Vol 1. Department of Information Science, Bergen

Flora, J. L. (1998). Social capital and communities of place. *Rural Sociology, 63*(4), 481–635.

Gellner, E. (1983). *Nations and nationalism*. Blackwell.

Gilchrist, A. (2009). *The well-connected community: A networking approach to community development*. Policy Press.

Go, J., & Watson, J. (2019). Anticolonial nationalism from imagined communities to colonial conflict. *European Journal of Sociology/Archives Européennes de Sociologie, 60*(1), 31–68.

Hargreaves, A. (2004). Building communities of place: Habitual movement around significant places. *Journal of Housing and the Built Environment, 19*(1), 45–65.

Harrington, C., Curtis, A., & Black, R. (2008). Locating communities in natural resource management. *Journal of Environmental Policy and Planning, 10*(2), 199–215.

Harvey, D. (1997, Winter/Spring). The new urbanism and the communitarian trap. *Harvard Design Magazine*, 68–71.

Head, B. W. (2007). Community engagement: Participation on whose terms? *Australian Journal of Political Science, 42*(3), 441–454.

Helm, C., & Mier, M. (2019). On the efficient market diffusion of intermittent renewable energies. *Energy Economics, 80*, 812–830.

Iazzolino, G., Sorrentino, N., Menniti, D., Pinnarelli, A., De Carolis, M., & Mendicino, L. (2022). Energy communities and key features emerged from business models review. *Energy Policy, 165*, 112929.

IRENA. (2020). *Stimulating investment in community energy: Broadening the ownership of renewables*. Retrieved December 11, 2023, from https://www.irena.org/publications/2020/Dec/Stimulating-investment-in-community-energy-Broadening-the-ownership-of-renewables

Itzigsohn, J., & vom Hau, M. (2006). Unfinished imagined communities: States, social movements, and nationalism in Latin America. *Theory and Society, 35*, 193–212.

Kelly, L. A. (1996). Race and place: Geographic and transcendent community in the post-Shaw era. *Vanderbilt Law Review, 49*(2), 227.

Khotari, U. (2001). Power, knowledge and social control in participatory development. In B. Cooke & D. Kothari (Eds.), *Participation: The new tyranny?* (pp. 139–152). Zed Books.

Kumar, A., & Aiken, G. T. (2020). A postcolonial critique of community energy: Searching for community as solidarity in India and Scotland. *Antipode, 53*(1), 200–221.

Lazdins, R., Mutule, A., & Zalostiba, D. (2021). PV energy communities—Challenges and barriers from a consumer perspective: A literature review. *Energies, 14*(16), 4873.

Lee, D., & Newby, H. (1983). *The problem of sociology: An introduction to the discipline*. Unwin Hyman.

Liljenfeldt, J., & Soares, J. (2020). *Developing community energy initiatives: A literature review for the project Local labour market in the energy transition*. Retrieved December 11, 2023, from http://uu.diva-portal.org/smash/get/diva2:1608810/FULLTEXT01.pdf

Mah, A. (2012). *Industrial ruination, community, and place: Landscapes and legacies of urban decline*. University of Toronto Press.

Malešević, S. (2020). Imagined communities and imaginary plots: Nationalisms, conspiracies, and pandemics in the Longue Durée. *Nationalities Papers, 50*, 45–60.

Mercer, D., & Marden, P. (2006). Ecologically sustainable development in a 'quarry' economy: One step forward, two steps back. *Geographical Research, 44*(2), 183–203.

Miller, H. T., & Demir, T. (2007). Policy communities. In S. F. Fischer & G. Miller (Eds.), *Handbook of public policy analysis: Theory, politics, and methods*. CRC Press.

Nash, V., & Christie, I. (2003). *Making sense of community*. Institute for Public Policy Research.

Nelson, N., & Wright, S. (1995). Participation and power. In N. Nelson & S. Wright (Eds.), *Power and participatory development: Theory and practice* (pp. 1–12). Intermediate Technology Publications.

Newman, L., & Dale, A. (2005). Network structure, diversity, and proactive resilience building: A response to Tompkins and Adger. *Ecology and Society, 10*(1). https://doi.org/10.5751/ES-01396-1001r02

Ostrom, E. (1990). *Governing the commons: The evolution of institutions for collective action*. Cambridge University Press.

Ostrom, E. (2005). *Understanding institutional diversity*. Princeton University Press.

Practical Action Policy Paper. (2016). *Poor people's energy outlook 2016*. Retrieved December 11, 2023, from https://practicalaction.org/knowledge-centre/resources/poor-peoples-energy-outlook-2016/

Rigon, A., & Castán Broto, V. (2021). *Inclusive urban development in the global south: Intersectionalities, inequities, and community*. Routledge.

Robertson, D., Smyth, J., & McIntosh, I. (2008). *Neighbourhood identity: People, time and place*. Joseph Rowntree Foundation.

Rosenfeld, H., & Duchin, M. (2022). Explainer: Communities of interest. In *Political geometry: Rethinking redistricting in the US with math, law, and everything in between* (pp. 235–245). Springer.

Sarsembayev, A. (1999). Imagined communities: Kazak nationalism and Kazakification in the 1990s. *Central Asian Survey, 18*(3), 319–346.

Shucksmith, M. (2000). Endogenous development, social capital, and social inclusion: Perspectives from LEADER in the UK. *Sociologia Ruralis, 40*(2), 208–218.

Singleton, S. (2000). Co-operation or capture? The paradox of co-management and community participation in natural resource management and environmental policy-making. *Environmental Politics, 9*(2), 1–21.

Skerrat, S., & Steiner, A. (2013). Working with communities of place: Complexities of empowerment. *Local Economy, 28*(3), 320–338.

Smith, A. D. (1991). *National identity*. Penguin.

Smith, R. M. (2004). Identities, interests, and the future of political science. *Perspectives on Politics, 2*(2), 301–312.

Staeheli, L. A. (2008). Citizenship and the problem of community. *Political Geography, 27*(1), 5–21.

Star, S. L. (1989). The structure of ill-structured solutions: Boundary objects and heterogeneous distributed problem solving. *Distributed Artificial Intelligence, 2*, 387–420.

Stephens, C. A. (2013). *The persistence of nationalism: From imagined communities to urban encounters*. Routledge.

Stratford, E., Armstrong, D., & Jaskolski, M. (2003) Relational spaces and the geopolitics of community participation in two Tasmanian local governments: A case for agonistic pluralism? *Transactions of the Institute of British Geographers, NS28*(4), 461–472.

Taarup-Esbensen, J. (2019). Making sense of risk—A sociological perspective on the management of risk. *Risk Analysis, 39*(4), 749–760.

Taylor, M., & Singleton, S. (1993). The communal resource: Transaction costs and the solution of collective action problems. *Politics and Society, 21*(2), 195–214.

Tönnies, F., & Loomis, C. P. (1957). *Community and Society* (Gemeinschaft und Gesellschaft). Michigan State University Press.

UN. (2021). *Leveraging energy access for advancing sustainable energy goals*. Retrieved December 11, 2023, from https://sdgs.un.org/publications/report-2021-sdg7-tag-policy-briefs-leveraging-energy-action-advancing-sustainable

Unnikrishnan, H., Gerullis, M. K., Cox, M., & Nagendra, H. (2023). Unpacking dynamics of diverse nested resource systems through a diagnostic approach. *Sustainability Science, 18*, 153–180.

Wenger, E. (1998). Communities of practice: Learning as a social system. *Systems Thinker, 9*(5), 2–3.

Wilkinson, K. P. (1991). *The community in rural America*. Greenwood Press.

Yael, T. (1995). The enigma of nationalism. *World Politics, 47*(3), 418–440.

CHAPTER 4

Gender Equality and Social Inclusion in Community Energy: An Intersectional Perspective

Vanesa Castán Broto

Abstract Gender has long been a central topic for discussion in relation to energy provision and energy access, but it becomes particularly prominent when considering community energy and transitions to sustainable energy. Gender relations continue to shape infrastructure networks. There is a gender gap in the energy transition both in terms of women and gender non-conforming people being able to access the green labour market, and in terms of understanding how gender relations are at the root of energy injustices. This chapter engages with feminist theory and asks what does an intersectional perspective mean for community energy. In particular, the chapter argues that an intersectional perspective enables a radical rethinking of the technological, political, social, and institutional dimensions of the energy transition.

Keywords Gender and energy · Feminist political ecology · Intersectionality · Black feminism

V. Castán Broto (✉)
University of Sheffield, Sheffield, UK
e-mail: v.castanbroto@sheffield.ac.uk

V. Castán Broto (ed.), *Community Energy and Sustainable Energy Transitions*, https://doi.org/10.1007/978-3-031-57938-7_4

4.1 Introduction

This chapter engages with gender equality and social inclusion as central aspects of community energy. Chapter 1 presented community energy as a tool to accelerate a just transition to sustainable energy. Two characteristics make this possible. First, community energy can act as a mechanism for social change, bringing together communities to gain control and autonomy over their energy supply. Second, community energy can identify differential vulnerabilities and capacities, helping deliver inclusive outcomes in energy access and improved resilience. However, community energy is not a neutral tool, and it does not advance social inclusion objectives on its own. Implementing community energy projects in a just way requires careful consideration of how existing structures of oppression, discrimination, and exclusion may be reproduced through new projects and dreams of community futures.

A just transition creates opportunities for everyone, distributes impacts fairly, and follows a negotiated approach to decision-making at all levels. Transitions, however, start from the point of entrenched social inequalities. They often fail to challenge existing inequities, and in the worst cases, they may reinforce them. For example, all countries face a gender gap in the energy sector that directly impacts the workforce and decision-making for energy-related decisions that impact everyone (Clancy & Feenstra, 2019). This gender gap hinders the transition to sustainable energy regarding who can be included in the process and how transition outcomes impact different populations. Other signifiers of inequality, such as nation, race, class, or sexuality, may also shape the outcomes of transitions, but to date, they have received limited attention in the literature (Newell, 2021).

This chapter proposes a feminist perspective on community energy to tackle the challenges of gender equality head-on. First, the chapter examines how gender inequalities manifest in the energy transition as a point of entry to examine the question of difference. Second, the chapter argues for expanding the concern with gender to an intersectional perspective that tackles multiple forms of inequality, engaging with the theoretical implications of such a move. Third, the chapter explores what it means to develop an intersectional perspective on community energy. Finally, the chapter ends with a reflection on the strategies used within the CESET project to facilitate research practices that put an intersectional perspective on community energy at its core.

4.2 Feminist Perspectives on Energy Transitions

Debates on energy policy have emphasised a notion of energy access as transformative, not only providing economic benefits but also redefining social structures and gender relations. As interests have moved towards a transition to ensure universal access to renewable energy, the emphasis on gender has become more prominent, emphasising the active role that women can play in activating positive change in the energy sector. For example, the Gender and Energy Compact, developed by a coalition of public, private, and civil society actors under the auspices of the United Nations, aims "to catalyse action towards gender equality and women's empowerment to accelerate a just, inclusive and sustainable energy transition."[1] The Compact highlights measures to facilitate women's empowerment, to lead and participate in the transition, and to gain control (not just access) over sustainable energy (Table 4.1). According to the Compact, this requires more projects putting women at their centre, changes in strategies and policies, access to finance and business services, career development measures, and higher-quality knowledge, for example, by collecting gender-disaggregated data.

The Compact is innovative in its treatment of the energy transition through a gender lens. It resonates with the concern that the energy transition cannot be assumed to deliver universal benefits and that its benefits and impacts are differentiated by, among other factors, gender (Rewald, 2017; Tjørring, 2016). Initiatives like the Compact also recast women away from the notion of hapless victims (Buckingham & Le Masson, 2017). Instead, women's empowerment, especially in terms of access and control over resources, has become the main strategy to address gender issues in energy policy and practice (Winther et al., 2017). For example, understanding how women entrepreneurs benefit from electricity (because women tend to lead smaller and more frugal enterprises) may inform strategies to maximise those benefits (Pueyo & Maestre, 2019). Women are also considered active participants, playing specific roles leading the transition (Allen et al., 2019), and facilitating innovation (Anditi et al., 2022; Baruah, 2015). A gender lens is presented as a condition for successful renewable energy projects (Gray et al.,

[1] https://www.un.org/en/energy-compacts/page/energy-compact-gender-equality-and-women%E2%80%99s-empowerment-sustainable-energy, retrieved January 16, 2024.

Table 4.1 Summary of the multi-stakeholder Gender and Energy Compact launched in 2022

Goal 1: Women have equal opportunity to lead, participate in, and benefit from a just, sustainable, and inclusive energy transition
Goal 2: Women have equal access to and control over sustainable energy products and services

Outcomes	*Examples of actions suggested in the compact*
Outcome 1: Energy and time poverty, as well as drudgery of women are eliminated by increasing women's access to and control over sustainable energy products and services	• Provide additional funding to projects that address explicitly the needs and views of women • Additional investment in energy-efficient appliances and end-uses that directly meet women's energy demands and reduce their unpaid care work
Outcome 2: Countries and regions (re)formulate and adopt more inclusive and gender-responsive energy access and transition pathways, strategies, and policies and adopt them	• Create spaces for the effective participation of women of different backgrounds in local energy policy-making, planning, implementation, and monitoring • Development of gender-responsive budgets
Outcome 3: Women-owned and -led businesses have increased access to resources, such as finance, sustainable energy, entrepreneurial capacity, and business development services	• Actions to increase the share of women entrepreneurs, women-led and women-owned businesses in the sustainable energy value chains • Public and private investments in women-led and women-owned businesses
Outcome 4: Career entry and advancement avenues for women working on energy access and sustainable energy are created	• Measures for the provision of jobs, skills, education, and career opportunities for all women • Enhance businesses' gender strategies • Consider gender parity and balance
Outcome 5: Knowledge, mechanisms, tools, and sex-disaggregated data are more available and of higher-quality	• Collect gender-disaggregated data and analyse and publish knowledge products, case studies, toolkits, and success stories to enhance knowledge, increase the visibility of women and improve the availability of data on the gender-energy nexus

Compact available at https://genderenergycompact.org/assets/2022/06/1June2022_Multi-Stakeholder-Gender-and-Energy-Compact.pdf

2019). Gender inequality also hinders delivering energy justice (Feenstra & Özerol, 2021). The transition is presented as a dynamic process which requires women to play an active role. However, as the feminist literature has emphasised, portraying women in this way also shifts the burden to women in terms of reclaiming their space in the energy sector.

Women's access to green jobs in the energy transition is a concern, as women tend to be restricted to administrative jobs with little decision-making power (Winther et al., 2017). There is a gender performance gap whereby fewer women than men benefit from employment in the transition. The causes for this gap entail a mix of external and internal constraints, from the difficulties for women technicians to access the renewable energy sector to the cultural prejudices that may lead young women to study energy engineering (Pueyo & Maestre, 2019). Systematic reviews have shown that transitions to renewable energy have both positive and negative impacts on gender equality and that they tend to shift, rather than eliminate, inequalities, for example, reproducing the prejudices that prevent women from accessing the field of energy engineering (Johnson et al., 2020).

While the gender gaps are an empirical reality of the energy transition, the emphasis on women obscures the complex working of gender relations in the energy sector. Such debates mirror previous debates on gender and development that proposed moving beyond empowering women to challenge gender-based power relations that generate difference (Moser, 2017). The extractive and market processes involved in the energy sector have profound impacts on gender relations (Cielo & Sarzosa, 2018). A narrow focus on gender differences distracts attention from the forms of discrimination already inscribed in the category of gender. In India, for example, male operators of renewable energy face discrimination because their existence does not fit toxic notions of masculinity and gender binary logic (Pergetti, 2023). Their work is hidden and undervalued, downplaying the importance of this work and the skills required to deliver it (Pergetti, 2023). Discourses of 'petro-masculinity' that link misogynistic gender models to racist projects of nationhood and fossil fuel exploitation are inspiring new authoritarian social movements that hinder and slow down the just transition (Daggett, 2018). Inequities are embedded in the constitution of social roles in ways that manifest structurally but also through everyday interactions, assumptions of normality, and learnt behaviours.

Research and policy often present women as disadvantaged actors and uncritically emphasise the benefits of energy access without considering the broader social and political conditions in which such access happens (Rewald, 2017). Commonly held beliefs about gender in energy policy, such as ideas that link modernisation and productivity to improvements in quality of life and the focus on single technologies that would 'save' women, are, in fact, myths because they find no justification in practice. Such gender and energy myths shape policies and strategies as they are used to legitimate and instrumentalise specific technologies or projects, sometimes at the expense of gender equality outcomes (Listo, 2018). They also may help to justify the continued use of fossil fuels (Wilson, 2018).

The transition, however, calls for feminist perspectives that challenge structures of patriarchy and capitalism at the heart of fossil fuel cultures, which goes beyond a focus on gender differences (Bell et al., 2020). Feminist thinkers have tried to develop alternative perspectives by reframing problems, focusing on implementation and governance, and offering long-term visions and ideals (Cannon & Chu, 2021). Community energy may provide opportunities to activate such a transformative feminist perspective in practice. Still, it is also open to forms of appropriation that may reproduce the very gender relations it aims to challenge. The strategy here is to think of community energy through an intersectional perspective, which can capture the interconnected nature of multiple forms of discrimination and exclusion. At the same time, an intersectional perspective challenges the categories themselves that enable such discrimination. For example, gender and energy critiques focused on binary conceptions of gender and sex reinforce, rather than challenge, the conceptual structures that reproduce gender-related inequities (Fathallah & Pyakurel, 2020). The following section explains some of the insights from such an intersectional perspective, before exploring its implications for the energy transition and community energy.

4.3 Intersectionality: Potential and Challenges

Intersectionality provides a way of looking at the world by recognising the varied impacts that structures of power and control have on different subjectivities. Individuals connect with society and others through the signifiers of identity that shape who they are, but these same signifiers condition experiences and behaviours. Intersectionality calls to interrogate the combined impact of different forms of identification as they translate into situated, unique experiences. This means challenging the very categories used to name those experiences, their homogeneous application (e.g. there are many ways of being a woman) and the political affiliations they generate. To understand how the transition to sustainable energy reduces or exacerbates existing social inequalities, intersectionality is both a tool to reveal the dynamic and interconnected nature of inequalities and a means to foster radical social change:

> Intersectionality investigates how intersecting power relations influence social relations across diverse societies as well as individual experiences in everyday life. As an analytic tool, intersectionality views categories of race, class, gender, sexuality, class, nation, ethnicity and age – among others – as interrelated and mutually shaping one another. Intersectionality is a way of understanding and explaining complexity in the world, in people, and in human experiences. (Collins & Bilge, 2020: p. 29)

As a tool for revealing inequalities, intersectionality not only recognises particularity but also warns that the assumed universality of identity categories (women, black) can become a means of oppression. In her classic work on intersectionality, Crenshaw examined how the categorical separation of gender and race in antidiscrimination law failed groups at the margins, such as black women (Crenshaw, 1989). The term intersectionality has encapsulated a long-held demand from black feminism while expanding to embrace multiple social justice projects across academia (Collins, 2019). In energy studies, intersectionality is approached as a principle to understand identity as an intersection of multiple categories of difference that challenges entrenched categorisations—such as the homogeneous vulnerable woman, which dominates energy policy (Pergetti, 2023).

There is a risk that such an approach is translated into an additive formula of spatially or temporally rooted inequalities. First, by 'intersection' the emphasis is on how different identification categories interact with overlapping or conflicting effects (Cho et al., 2013). Second, intersectionality rejects categories that hold connotations of unity and stability and that, in doing so, simplify the experiences of inequality and reproduce an essentialist logic (Carastathis, 2013; Walby et al., 2012). Intersectionality is itself a critique of uncritical forms of categorisation. Third, intersectionality insists that the attention should be redirected towards the structures of power that create discrimination rather than mapping difference (Cho et al., 2013; Crenshaw, 1991). Categories of difference also matter insofar as they are able to articulate alliances and political coalitions to counter such projects of domination (Carastathis, 2013; Matsuda, 1990). In summary, intersectionality calls for a situated analysis of power structures as they translate into social categorisation, which in turn determines how individuals and groups relate to energy systems and their impacts (Kaijser & Kronsell, 2014).

This is why intersectionality emerges as much more than an analytical tool. A history of engagement with the lived political realities of marginalised people has made intersectionality a practical tool to achieve social change (Collins, 2019). In the context of the transition to sustainable energy, intersectionality highlights the interconnections between the structures of exploitation and the oppression and discrimination of groups of people on the basis of identity (Newell, 2021; Wilson, 2018). These intersecting forms of oppression call for flexible solidarities alongside forms of epistemic resistance (Collins, 2019). In the case of the transition to sustainable energy, this involves articulating collective claims to energy sovereignty, claiming the exploitation of resources, the technological systems that support them, and the forms of infrastructural violence they create (Castán Broto, 2017). Intersectionality raises 'who' questions: who acts, who receives, and how that leads to particular operations of political power in energy transitions (Ryder, 2018). The ultimate objective is not to navigate but to dismantle existing forms of discrimination.

4.4 An Intersectional Perspective on Community Energy

Intersectionality has become popular in many fields, but often, this has been by adapting concepts and methods to the new concerns raised by intersectionality rather than accepting intersectionality's fundamental challenge to existing conceptualisations of social life (Cho et al., 2013). This chapter aims to show how intersectionality helps rethink the transition to sustainable energy and the role of community energy, first, by positioning community energy as a strategy that puts social equality at the heart of the energy transition and, second, by exploring what an intersectionality perspective may mean for the deployment of community energy projects.

4.4.1 Intersectional Transitions and Community Energy

Identifying the structures of privilege that shape the transition to sustainable energy becomes the first analytical task for the intersectionality scholar (Garcia & Tschakert, 2022). In this case, the reproduction of a networked ideal camouflages patriarchal and colonial power under the pretence of universal service provision (Coutard, 2008; Graham & Marvin, 2002). Post-networked ideals are paradoxical because they advance a political critique of universalism while accepting the retreat of the state and the liberalisation of services (Coutard & Rutherford, 2015). Their rise corresponds to a parallel rise in the popularity of resilience ideals as situated and grounded, explicitly addressing power's workings (see Chapter 2). While off-grid systems offer hope, they effectively shift the burden of work to the disadvantaged populations which will putatively benefit from them (Castán Broto, 2022).

From a feminist perspective, distributed and decentralised fuel power challenges existing power structures (Bell et al., 2020). Decentralisation is a crucial pillar of a feminist perspective on the transition to sustainable energy together with a plural policy landscape, an impulse to center care for humans and the Earth, the rejection of any form of exploitation, and the scepticism towards technosolutionism (Table 4.2). The challenge is that empirical evidence does not follow directly from feminist theory. As explained above, research on inequalities in energy access has tended to use universal categories, fixating, for example, on the benefits of electrification or any other energy innovation to the lives of women, and they have hardly considered the impacts of different systems on different populations. Claims on the benefits of low-carbon or off-grid developments on the lives of disadvantaged groups (with an abstract category of women often standing for those) rarely present tangible evidence of local and immediate impacts (Rewald, 2017).

Table 4.2 shows how community energy projects respond to feminist ideals of the transition to sustainable energy. Community energy works when the community concerned mobilises to claim sovereignty over its energy resources. The community needs to activate a coalition-building process to mobilise flexible solidarities, in which the priority is not to leave anyone behind. Notions of plurality and dialogue are coupled with the realities of technology access for communities to claim autonomy over

Table 4.2 Dimensions of a feminist research agenda on the transition to sustainable energy

Dimension	*Challenge*	*Vision for the transition*	*Implications for community energy*
Political	"Renewable energy systems do not automatically produce democracy and justice, nor are they necessarily sustainable. New fuel technologies alone, without a matching new fuel politics, are unlikely to resolve the looming climate disaster"	Democratic, decolonial, decentralised, pluralist, publicly-owned	Community energy projects facilitate the inclusion of a plurality of perspectives in their configuration. This may require active attempts to facilitate the inclusion of those social groups marginalised from the community or in the margins and the construction of open arenas of deliberation where all members feel heard and included
Economic	"Resist the ascetic framing of energy consumption—in which only economic growth, based upon intensive energy consumption, produces well-being—and build upon feminist visions of community thriving and pleasure predicated upon alternative energy systems"	Prioritises human and more-than-human well-being and biodiversity over profit; refuses the growth imperative; committed to community economies and pink-collar jobs	Community energy projects prioritise community well-being and livelihood opportunities over the commercialisation of electricity beyond the community. The objective is for the community to maintain sovereignty over their energy resources

(continued)

Table 4.2 (continued)

Dimension	*Challenge*	*Vision for the transition*	*Implications for community energy*
Socio-ecological	"Energy systems, like all human terraforming projects, are inherently violent and unjust. But waiting for purity results in paralysis and serves the interests of the carbon-emitting status quo"	It is relational, transparent, attuned to the violence of energy production and committed to building a culture of care	Community energy projects acknowledge the forms of exploitation inherent in energy production and distribution through, for example, the process of sitting the project
Technological	"Technology is not politically neutral and should not be controlled by the highest bidders, who are almost always transnational corporations that emphasise profit over people"	Distributed, community-directed and collaborative, heterogeneous and multiple	Community energy advances technology justice, recognising the knowledge and technical skills within a community and their right to access more beneficial technologies from elsewhere

Adapted from Bell et al. (2020)

their present and future energy (Castán Broto, 2017). Such a unique process can be transformative, but it also poses complex questions of inclusion within every community energy project as it unfolds.

4.4.2 *Making Community Energy Work for Everyone*

In practice, intersectionality requires situating any categorisations of identity in their historical and geographical contexts and not treating any single category in isolation. Gender, or any other form of identity, should also not be treated as synonymous with marginalisation. Any gender projects should recognise instead the locations of privilege and the dynamics of discrimination. For community energy projects, this means engaging actively with the process whereby a community comes together in the form of a coalition and how they advance particular projects of self-improvement, in this case in the energy sector, remembering that no project is intrinsically positive for everyone (Ojong, 2021). This does not mean, however, overlooking the constraints faced by social groups facing discrimination. Women, for example, face bundled constraints in accessing labour markets in renewable energy, which directly impact their participation as active agents in community energy projects (Table 4.3). A community energy project may start by reviewing how these bundled constraints impact women, but it cannot do so without asking 'the other question' (Carastathis, 2013; Matsuda, 1990), for example, who are the people who face barriers to access skills and education and why?

A community energy project will require co-production to bring together different actors, challenge their assumptions, and develop collective goals. Castán Broto and Neves Alves (2018) developed an analytical tool to think of intersectionality in relation to the co-production outcomes that community energy projects may aspire to (Table 4.4). The outcomes of co-production vary, ranging from the production of a design, the facilitation of institutional innovation, the development of new inclusive decision-making processes, and the creation of new systems of signification. For example, a community energy project will produce a context-specific design, new institutions to manage and maintain the project that include the community, a decision-making process that puts the community at its centre, and new ideas about energy access and the kinds of dependencies that reproduce inequalities in energy access. These are all overlapping stages in which the community will have a say. For the communities to have such a say, they need to be recognised as legitimate interlocutors of the project. That, however, is not always easy (see Chapter 3). First, there is a question of what the community is and who belongs to it. Second, there is a question about how the community is perceived from the outside. Third, there is a question about the extent to which the community is recognized as a legitimate interlocutor by various actors. Fourth, there is a question of the inherent deficit of credibility that some people face by virtue of their position in society (this deficit of credibility is called epistemic injustice). Each of these elements poses a specific recognition challenge, to which intersectionality approaches provide a range of insights.

In summary, the operationalisation of intersectionality will vary across implementation contexts, but three principles may help start the analysis:

- To recognize people at the receiving end of diverse forms of institutional, structural, and cultural discrimination that prevent their flourishing and

Table 4.3 Addressing the bundled constraints faced by women through community energy systems

Bundled constraints that women face	*Implications for women's livelihoods*	*Implications for the incorporation of women in community energy projects*
Access to skills and education	More likely to depend on low or subsistence wages Less security Limited access to decent work	Additional training needs may be required
Low access and control over resources such as land, inputs, energy, or income	Limited access to capital Not being able to purchase time-saving equipment	Formation of savings groups or cooperatives may help women to accumulate capital over time
Unequal distribution of care responsibilities	Prioritising flexible patterns of work to accommodate care work Bearing the burden of unpaid care work	Adapting the rhythms of work in community energy projects to those of care
Restrictions on women's use of space or mobility	Social norms and expectations may influence what spaces women can access or not and at which times, including having different mobility needs	Adequate safety provisions need to be considered in the design and development of community energy projects
Capacity to influence	Women may be explicitly excluded from decision-making processes that affect them	Governance mechanisms need to incorporate measures to integrate women in decision-making, including forms of positive discrimination when needed
Occupational segregation	Women tend to gravitate towards sectors considered as 'female sectors', away from male ones	Female technicians are underrepresented in the renewable energy sector: additional training may support skills development

Extending an analysis of Pueyo and Maestre (2019)

- To recognize people as active agents in charge of their own futures rather than as passive receivers of the benefits of a given community energy intervention.
- To direct action towards dismantling structural barriers to emancipation and making it possible for people to use their capacities to

Table 4.4 Intersectionality dimensions of service co-production

Outcomes	*Recognition challenges*	*Intersectionality insights*
Context-situated technological and co-produced design	Whose uses are prioritized in the design of a project? Whose values are taken into consideration when developing context-based solutions and technologies? What uses and needs are constantly overlooked?	Single categories for social analysis (e.g., gender, age, income) do not reflect the needs of diverse groups Access to services depends more broadly on individuals' position in the household, the community, and society. It is also shaped by the everyday practices of different social groups. Focusing on these everyday practices can help understand the mechanisms and power dynamics shaping uneven access to services, and to assess how co-design can help overcome this
Institutional innovation and collective organization of service provision	Whose needs are prioritized in service delivery? How do existing systems of provision serve different groups? What capacity do those groups have to participate in institutions for service provision?	Co-production may disrupt existing social categories that give power to certain actors over others. These disruptions may also affect social strategies that depend on people's changing individual identities in multiple situational contexts as well as the changing social categories and groups they belong to in equally diverse situations—co-production processes are part of subjectivity-making processes

Outcomes	*Recognition challenges*	*Intersectionality insights*
New and inclusive processes of decision-making, planning, and urban governance	Who can access decision-making processes, and in what terms? Who is excluded and how?	The specific conditions of decision-making already have consequences for the inclusion and exclusion of particular social groups. Requirements to participate in the decision-making processes may pose an additional burden to comparatively vulnerable groups if the conditions for participation are not favourable. Rights-based approaches may contribute to the exclusion of people who suffer forms of oppression that cannot be reflected in a streamlined framework
New systems of signification, change of paradigms	What perspectives on reality create instances of symbolic violence and reproduce existing forms of oppression and exclusion?	The principles of patriarchy, racism and colonialism continue to be reproduced through well-intentioned practices of emancipation if they do not recognize experiences that are not reflected in well-established social categories. Attempts to claim the right to free oneself from oppressive structures may inadvertently reproduce inequality

Castán Broto and Neves Alves (2018)

improve their autonomy and flourishing and recognise the location of privilege.

4.5 Implementing an Intersectional Perspective in a Research Project on Community Energy

How to put equality and intersectionality at the core of CESET? Challenging inequality must become an ongoing process in which structural and cultural barriers are challenged through a centripetal intersectional practice (Cho et al., 2013): one that decentres the assumptions and philosophy of the project. This means mobilising a concern with inequality in every structural decision about the project, but also in day-to-day behaviours and attitudes. For CESET, intersectionality is a diagnostic tool to analyse the drivers of inequity, support civil struggles, and prevent unnecessary harm. Paraphrasing the Combahee River Collective Statement (1977), a key objective of the project is 'not to mess with people' in the name of externally imposed project goals. While these aspirations are laudable, there are many challenges to delivering them in practice. Perhaps the most salient is that CESET researchers are not members of the communities where energy projects are being developed, although over time some of those members have been brought into dialogue with the team (see, as an example, Chapters 10 and 11).

CESET aims to recognize the drivers of inequality (using the project to challenge them) and mobilise existing capacities and possibilities that support emancipatory directions. The off-grid energy lab (CESET's Work Package 4) will be implemented by the Universidade Eduardo Mondlane together with CESET's delivery partner, SCENE and the University of Sheffield. The objective is to create a laboratory to build local capacities for understanding and delivering renewable energy. We are also investigating how the project could provide energy services to a local community (e.g., energy for a community hub, street lighting, or directly to households). CESET's plan is to develop an off-grid, flexible system that could grow organically through the support and collaboration of residents. Early on, it was deemed that women's needs were underrepresented in Mozambique's energy policy, and thus, the project developed policy recommendations that brought the team into a dialogue with the Ministry of Mineral Resources and Energy (MIREME), dring their development of a gender strategy (Table 4.5). In doing so, the project has fostered a cross-institutional dialogue on gender and energy: a starting point for further policy development.

However, these suggestions alone are not sufficient to deliver the coalitions that will deliver just energy through community energy projects. The category of 'women' (or even 'women and other') presents its own challenges. The lab will be implemented in the neighbourhood of Chamanculo, in Maputo. Multiple factors, from origin to political affiliation, shape the power structures of the neighbourhood alongside more conventional class and gender lines. The project needs to engage with such power structures while also building bridges to mobilise the flexible solidarities that support emancipatory coalitions.

For this reason, CESET's approach cannot rely solely on a set of policy recommendations or prescriptions. The point is to move from aspirational utopias and engage with practical actions on the ground, with the kind of care already anticipated in black feminist thought. Three challenges that conflict with the delivery of community energy are:

- The centrality of ideas of care and mutual support to enable the participation of all kinds of people while also understanding how certain groups carry a disproportionate burden of care work.
- The need to demystify technologies and make them accessible to everyone.
- The importance of breaking with epistemic injustice and giving credibility to people's experiences.

A feasible approach is one accepting of the dynamic and incomplete process of making an alliance, in which the most important challenge is to keep checking in. A starting point could be to reflect on the development of the project. For that task, Castán Broto and Neves Alves (2018) provide a set of questions that enables a continuous and generous engagement with the minutiae of building political and technological power at the community level (Table 4.4). Table 4.6 reimagines those questions in the case of community energy. These questions have been discussed across the project team and with community leaders, to facilitate local leadership in the implementation of CESET's energy lab. As the project is being delivered in the spring of 2024, the team is looking forward to see what transformations will it foster.

So, this chapter concludes as an open process of rethinking and reimagination: Without strong prescriptions for action. Instead, the chapter provides an invitation to embrace complexity as a constitutive factor of

Table 4.5 Recommendations to advance gender equality objectives in CESET

Recommendation	*Implications*
Engage women along different segments of the off-grid renewable energy value chain	A gender perspective needs to be integrated from the very beginning in the design, implementation, and monitoring of energy access programmes. Women need to be seen as key actors in the delivery of energy solutions and not just as primary end-users and beneficiaries of such programmes
Recognize the specific set of challenges and opportunities faced by women in energy access contexts	Off-grid renewable energy solutions promise substantial improvements in women's quality of life through reduced indoor air pollution, better lighting to enable girls to study, less day-to-day drudgery and more opportunities to earn income from productive uses of energy or from the jobs that modern energy induces in rural economies
Ensure women's participation in planning	Given women's role as primary energy users in the household, their participation in planning and implementing a sustainable renewable energy system is essential if their priorities and preferences are to be reflected in the system
Involve women in project delivery	As women become engaged in delivering energy solutions, they take on more active roles in their communities and consequently facilitate a gradual shift in the social and cultural norms that previously acted as barriers to their agency
Support skills development	Support skills development in technical and non-technical subjects and in broad business and leadership skills and emphasize the need to engage women in design, delivery, and operations Some skills needed to install, operate, and maintain off-grid solutions require little past education or prior experience; thus can skills be developed locally and training for the technology can be done on-site or on the job
Contest the broader context insofar as actively promotes discrimination based on gender or other protected characteristics	Examine the cultural beliefs about the different roles of energy in people's lives and the way energy is used and examine the regulatory context of energy provision as a means to examine whether active discrimination is taking place

Adapted from IRENA (2019)

Table 4.6 Questions for the continuous evaluation of intersectionality concerns in the CESET lab

Critical questions	
Whose uses are prioritized in the design of our activities? Whose values are taken into consideration when developing context-based solutions and technologies? What uses and needs are constantly overlooked?	Co-producing an off-grid energy project with a local community in Maputo will require understanding different energy uses within the neighbourhood and how they relate to people's perceptions of their energy needs. Involving diverse potential users in the early stages of research through surveys and informal contacts will be essential The team will also need to reflect on how working with official authorities (e.g., municipality) will affect different groups and individuals' willingness/capacity to take part in the project. The research team may have biases that may also influence whose needs are prioritized
Whose services are prioritized? How do existing systems of provision serve different groups? What capacity do those groups have to participate in institutions for service provision?	The emphasis of the laboratory is on collective services, but the team will need to reflect on who is defining what important collective usage might be and what collective services are needed. The structures of governance at the neighbourhood level will influence CESET's perceptions of needs, because the local leaders involved in the project might not represent the whole community. Thus, it will be important to create opportunities to discuss those needs openly
Who can access decision-making processes, and in what terms? Who is excluded and how?	CESET must create the possibility to hear a diversity of voices through a diversity of means. The aspiration is to create non-confrontational, open processes in which different people feel sufficiently safe to speak their mind. Previous experiences in Maputo suggest that while energy is a difficult topic, emphasizing the enjoyment of being together and discussing common problems may be a means to bring together different groups. This also means paying attention to the role that local leaders play redirecting the discussion and possibly silencing some voices

(continued)

Table 4.6 (continued)

Critical questions	
What perspectives create instances of symbolic violence and reproduce existing forms of oppression and exclusion?	This question raises concerns about community energy as a unique solution and how it meets the needs. The project may become a project for its own sake, without really taking into consideration what the energy needs of the community are. Making sure that diverse community perspectives are represented in the project is essential. Also, understanding the kind of energy futures that are desired by communities is of vital importance, particularly if the community energy project risks displacing alternative livelihoods or means of understanding the city

everyday life in which justice is always an objective in the horizon capable of inspiring shared commitments across perspectives and understandings.

References

Allen, E., Lyons, H., & Stephens, J. C. (2019). Women's leadership in renewable transformation, energy justice and energy democracy: Redistributing power. *Energy Research & Social Science, 57*, 101233.

Anditi, C., Musango, J. K., Smit, S., & Ceschin, F. (2022). Addressing gender dimensions in energy innovations: A gender analysis framework for informal urban settlements in Africa. *Energy Research and Social Science, 88*, 102476.

Baruah, B. (2015). Creating opportunities for women in the renewable energy sector: Findings from India. *Feminist Economics, 21*(2), 53–76.

Bell, S. E., Daggett, C., & Labuski, C. (2020). Toward feminist energy systems: Why adding women and solar panels is not enough. *Energy Research & Social Science, 68*, 101557.

Buckingham, S., & Le Masson, V. (2017). *Understanding climate change through gender relations*. Taylor & Francis.

Cannon, C. E. B., & Chu, E. K. (2021). Gender, sexuality, and feminist critiques in energy research: A review and call for transversal thinking. *Energy Research & Social Science, 75*, 102005.

Carastathis, A. (2013). Identity categories as potential coalitions. *Signs: Journal of Women in Culture and Society, 38*(4), 941–965.

Castán Broto, V. (2017). Energy sovereignty and development planning: The case of Maputo, Mozambique. *International Development Planning Review, 39*(3), 229–248.

Castán Broto, V. (2022). Splintering urbanism and climate breakdown. *Journal of Urban Technology, 29*(1), 87–93.

Castán Broto, V., & Neves Alves, S. (2018). Intersectionality challenges for the co-production of urban services: Notes for a theoretical and methodological agenda. *Environment and Urbanization, 30*(2), 367–386.

Cho, S., Crenshaw, K. W., & McCall, L. (2013). Toward a field of intersectionality studies: Theory, applications, and praxis. *Signs: Journal of Women in Culture and Society, 38*(4), 785–810.

Cielo, C., & Sarzosa, N. C. (2018). Transformed territories of gendered care work in Ecuador's petroleum circuit. *Conservation and Society, 16*(1), 8–20.

Clancy, J., & Feenstra, M. (2019). *Women, gender equality and the energy transition in the EU*. Publications Office of the European Union.

Collins, P. H. (2019). *Intersectionality as critical social theory*. Duke University Press.

Collins, P. H., & Bilge, S. (2020). *Intersectionality*. Wiley.

Combahee River Collective. (1977). *The Combahee river collective statement*. Retrieved January 9, 2024, from https://www.blackpast.org/african-american-history/combahee-river-collective-statement-1977/

Coutard, O. (2008). Placing splintering urbanism: Introduction. *Geoforum, 39*(6), 1815–1820.

Coutard, O., & Rutherford, J. (2015). *Beyond the networked city: Infrastructure reconfigurations and urban change in the North and South*. Routledge.

Crenshaw, K. (1989). Demarginalizing the intersection of race and sex: A black feminist critique of antidiscrimination doctrine, feminist theory and antiracist politics. *University of Chicago Legal Forum, 1989*, 139–167.

Crenshaw, K. (1991). Mapping the margins: Intersectionality, identity politics, and violence against women of color. *Stanford Law Review, 43*(6), 1241–1299.

Daggett, C. (2018). Petro-masculinity: Fossil fuels and authoritarian desire. *Millennium, 47*(1), 25–44.

Fathallah, J., & Pyakurel, P. (2020). Addressing gender in energy studies. *Energy Research & Social Science, 65*, 101461.

Feenstra, M., & Özerol, G. (2021). Energy justice as a search light for gender-energy nexus: Towards a conceptual framework. *Renewable and Sustainable Energy Reviews, 138*, 110668.

Garcia, A., & Tschakert, P. (2022). Intersectional subjectivities and climate change adaptation: An attentive analytical approach for examining power, emancipatory processes, and transformation. *Transactions of the Institute of British Geographers, 47*(3), 651–665.

Graham, S., & Marvin, S. (2002). *Splintering urbanism: Networked infrastructures, technological mobilities and the urban condition*. Routledge.

Gray, L., Boyle, A., Francks, E., & Yu, V. (2019). The power of small-scale solar: Gender, energy poverty, and entrepreneurship in Tanzania. *Development in Practice, 29*(1), 26–39.

IRENA. (2019). *Renewable Energy: A Gender Perspective*. International Renewable Energy Agency.

Johnson, O. W., Han, J.Y.-C., Knight, A.-L., Mortensen, S., Aung, M. T., Boyland, M., & Resurrección, B. P. (2020). Intersectionality and energy transitions: A review of gender, social equity and low-carbon energy. *Energy Research & Social Science, 70*, 101774.

Kaijser, A., & Kronsell, A. (2014). Climate change through the lens of intersectionality. *Environmental Politics, 23*(3), 417–433.

Listo, R. (2018). Gender myths in energy poverty literature: A critical discourse analysis. *Energy Research & Social Science, 38*, 9–18.

Matsuda, M. J. (1990). Beside my sister, facing the enemy: Legal theory out of coalition. *Stanford Law Review, 43*, 1183.

Moser, C. O. N. (2017). Gender transformation in a new global urban agenda: Challenges for Habitat III and beyond. *Environment and Urbanization, 29*(1), 221–236.

Newell, P. (2021). Race and the politics of energy transitions. *Energy Research & Social Science, 71*, 101839.

Ojong, N. (2021). The rise of solar home systems in sub-Saharan Africa: Examining gender, class, and sustainability. *Energy Research & Social Science, 75*, 102011.

Pergetti, S. (2023). Sustainability is not a thing: It's hard work! Recognizing the middlemen of operations and maintenance for feminist energy systems in India's Sundarbans. *Energy Research & Social Science, 105*, 103290.

Pueyo, A., & Maestre, M. (2019). Linking energy access, gender and poverty: A review of the literature on productive uses of energy. *Energy Research & Social Science, 53*, 170–181.

Rewald, R. (2017). *Energy and women and girls: Analyzing the needs, uses, and impacts of energy on women and girls in the developing world*. Oxfam Research Backgrounder Series. Retrieved January 9, 2024, from https://www.oxfamamerica.org/explore/research-publications/energy-women-girls

Ryder, S. S. (2018). Developing an intersectionally-informed, multi-sited, critical policy ethnography to examine power and procedural justice in multiscalar energy and climate change decisionmaking processes. *Energy Research & Social Science, 45*, 266–275.

Tjørring, L. (2016). We forgot half of the population! The significance of gender in Danish energy renovation projects. *Energy Research & Social Science, 22*, 115–124.

Walby, S., Armstrong, J., & Strid, S. (2012). Intersectionality: Multiple inequalities in social theory. *Sociology, 46*(2), 224–240.
Wilson, S. (2018). Energy imaginaries: Feminist and decolonial futures. *Materialism and the Critique of Energy, 31*(2), 5–20.
Winther, T., Matinga, M. N., Ulsrud, K., & Standal, K. (2017). Women's empowerment through electricity access: Scoping study and proposal for a framework of analysis. *Journal of Development Effectiveness, 9*(3), 389–417.

CHAPTER 5

Financial Aspects of Community Energy Systems

Aran Eales

Abstract This chapter analyses the process of financing community energy, emphasising delivering sustainable projects that do not add additional burdens for communities. Community energy depends on developing micro- and pico-systems, which have specific financial requirements due to their small size. A key challenge is ensuring the financial sustainability of projects once installed and when the initial capital runs out. The chapter advocates putting the community at the centre of a structured financial planning process, negotiating cost reductions and revenue models, and diversifying the revenue stream by looking beyond conventional sources of finance. The analysis of a case study of financing two micro-grids in Malawi demonstrates the obstacles faced by financing projects. These small projects face enormous challenges that drive up costs, including difficulties in accessing supply chains for solar equipment, inflation and foreign exchange fluctuations, and limited technical capacity to maintain the system. A sustainable energy transition should help reduce these costs to facilitate the expansion of the micro-grid model. Until such

A. Eales (✉)
University of Strathclyde, Glasgow, UK
e-mail: aran.eales@strath.ac.uk

V. Castán Broto (ed.), *Community Energy and Sustainable Energy Transitions*, https://doi.org/10.1007/978-3-031-57938-7_5

expansion takes place, the sustainability of micro-grids will depend on public support and external grants to top up the income from consumer tariffs.

Keywords Community energy finance · Capital costs · Operational costs · Supply chains

5.1 Introduction

Community Energy Systems (CES) play a pivotal role in providing sustainable and affordable electricity in off-grid communities. Yet, a comprehensive understanding of their financial landscape is crucial for successful implementation. This chapter aims to shed light on economic considerations that shape the viability and sustainability of CES, with a specific emphasis on micro- and pico-grid systems (1 to 50 kW) operating commercially in rural areas of the Global South. Given their smaller size, such systems will have different requirements in terms of resources, materials, and capacity than larger or grid-connected systems and may depend on different funding mechanisms.

The financial management of CES has garnered increasing attention as a pivotal aspect of sustainable and decentralised energy solutions, particularly in the context of mini-grids and off-grid systems. Several high-level industry reports have provided valuable insights into the global state of play, market trends, potential investment opportunities within this sector, and trends in technology costs and business models. While industry reports offer a macroscopic view, detailed case studies with specific shared primary data on financial performance, crucial for a nuanced understanding, remain rare. An emerging body of academic literature assesses CES in terms of techno-economic perspectives and business modelling. Still, there is a lack of robust academic discourse that explores, systematically interrogates, and quantitatively assesses the business feasibility and financial management of CES. Ultimately, despite growing interest and increasing discourse, proven sustainable financial models are scarce for CES, and the specific landscape of CES financial models remains largely uncharted.

The chapter discusses some principles for the financial management of CES. It outlines steps for developing a financial plan for their sustainable deployment and operation, drawing on previous experiences with micro-grids. These steps broadly involve balancing capital and operating expenditure with revenue from selling electricity, considering characteristics unique or particularly relevant to CES. A key lesson is to ensure that the operational costs of CES are considered in financial planning. The chapter thus explains the main costs and sources of revenue for CES, as well as considering multiple sources of finance. A case of a CES in Malawi helps demonstrate how these principles apply in practice.

5.2 Principles for the Financial Management of CES

CES require upfront capital to construct and install the systems, including developing ancillary infrastructures to make the project viable. Sustaining the CES's functionality over the project life requires an additional, continuous, reliable revenue stream. Sufficient funds are needed for ongoing operations, maintenance, and the effective management of the systems (Safdar, 2017). This dual financial strategy ensures the successful implementation, longevity, and effectiveness of CES by addressing both its foundational development and sustained operational needs.

Off-grid renewable energy systems have historically encountered sustainability challenges. While donor capital has been deployed to develop energy infrastructure, the absence of a financially sustainable business model has frequently led to insufficient resources for maintenance or the replacement of components. This deficit in ongoing funding and adequate business models has, in turn, resulted in the deterioration of systems over time (Dauenhauer et al., 2019). Accordingly, a key guiding principle in the financial management of CES is to ensure sufficient resources are available to cover the costs of operation, maintenance, and management to ensure long-term sustainability. Such costs can be covered with revenues from connection fees and electricity sales and, where available, from subsidies or donor support. In any case, ensuring a reliable and ongoing source of revenue is vital to the project's sustainability (IRENA, 2018).

Another key guiding principle regarding the financial management of CES is to carry out a cost–benefit analysis for the project that assesses the relation between the cost of the proposed CES and the value of the resulting benefits, specifically to the community it serves. The benefits considered in such an assessment can be both tangible and intangible:

- **Tangible Benefits:** Direct, measurable advantages such as reduced energy costs for community members, increased economic activities, and job creation.
- **Intangible Benefits:** Less quantifiable yet impactful outcomes, including enhanced community cohesion, improved health outcomes, and environmental conservation.

Creating a robust financial plan for a CES requires a systematic and iterative approach. The process involves estimating costs, developing an initial revenue model, and testing it through community consultation so that the project aligns with the community's demand. This iterative process ensures flexibility and adaptability to unique community needs. Figure 5.1 outlines a typical process based on micro-grid literature (Weston et al., 2018). The model puts the community at the centre of financial planning.

The steps begin with a detailed estimation of the costs of implementing and operating the CES. This includes infrastructure, technology, personnel, and ongoing maintenance expenses. A revenue model is then developed based on projected energy demand and potential tariffs balanced with community affordability. Additional revenue streams, including grants, subsidies, or income-generating activities linked to the CES, can help diversify the revenue stream.

Engaging the community in the financial planning process is needed to align the revenue model with community needs, ensure community buy-in, and ultimately contribute to long-term sustainability. This is achieved

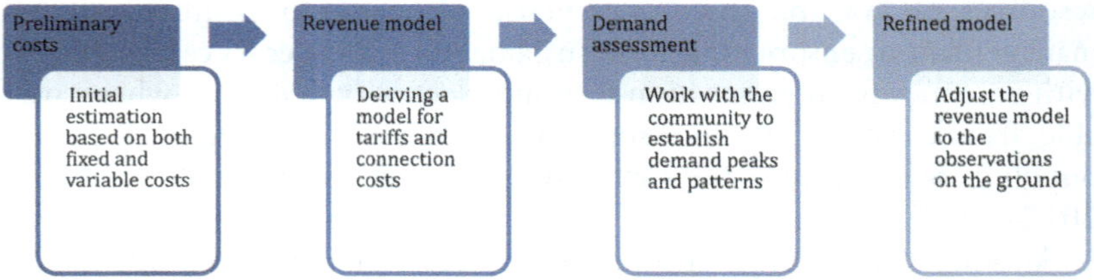

Fig. 5.1 Steps towards the development of a financial plan

through clearly understanding customer demand and seeking input on affordability, expectations, and potential contributions to the CES. The initial revenue model can then be tested against the community's feedback, analysing its feasibility and refining it based on community preferences to ensure it aligns with the overall objectives of energy access and community resilience.

Direct negotiation with the community may also help to reduce specific costs. This involves collaborative decision-making on maintenance, resource allocation, or shared responsibilities. Establishing a continuous review process to monitor the financial plan's performance, which regularly assesses whether the CES's financial objectives align with the evolving needs and dynamics of the community, is also required.

5.3 Costs of CESs

From the point of view of investment and financial management of a CES, it is helpful to distinguish between capital expenditure (CAPEX) costs and operating expenses (OPEX).

- CAPEX are the major investments that will take place during the project's life. In terms of investment, they include long-term capital expenditures (infrastructure and equipment) for purchases that will be used for longer than a year.
- In contrast, operating expenses (OPEX) are the expenses that are required to keep the infrastructure working, such as maintenance contracts, site staff wages, as well as business costs, including rent, transport, and overheads.

Fully understanding CAPEX and OPEX costs requires substantial stakeholder engagement, technical design iteration, financial modelling iteration, regulatory approvals, community governance, and developing sustainable operational models (Fig. 5.2). There can be a tendency to underestimate overheads, transaction costs and the management of customer relationships, which should be avoided through project planning.

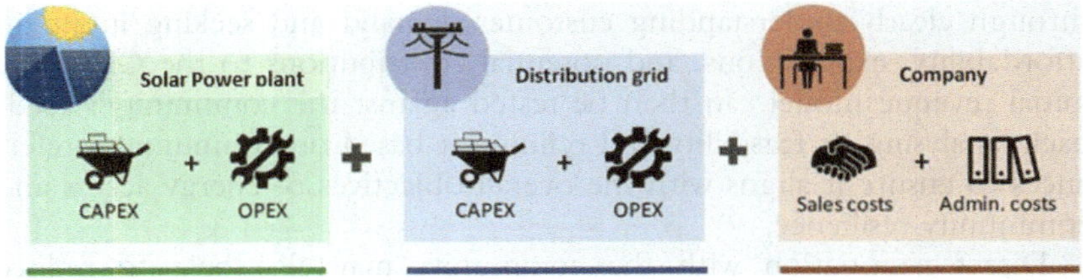

Fig. 5.2 Costs of installing and operating a solar micro-grid CES

5.3.1 Capital Expenditure (CAPEX)

A micro-grid generally comprises a generation unit (solar, wind, hydro, bioenergy, or diesel), distribution infrastructure (wires and poles for transporting the electricity to customer connections as well as premises wiring), customer consumption monitoring (meters or smart meters), and remote monitoring (AMDA, 2020).

Calculating CAPEX for a CES involves a systematic approach encompassing multiple key steps. Once a community has been identified through detailed site selection, factors such as geographical location and community needs are considered, and a comprehensive demand assessment is carried out. This is conducted through surveys or utilising measured data from analogous projects and provides insights into the energy requirements of the targeted community. Subsequently, the technical design phase involves sizing components for the generation and distribution aspects to meet the demand using the available renewable resources, which informs a detailed Bill of Quantities[1] with associated costs derived from local suppliers. Figure 5.3 shows a typical breakdown of Capex for a 30 kW solar-diesel hybrid micro-grid.

In addition to CAPEX for components for the CES, installation costs must be included taking into consideration wages, transport, and other overheads of the local installation team. Community engagement is another integral cost, evaluated in terms of organising awareness programmes, fostering local support, and ensuring the active involvement of community members through training and workshops. Other

[1] A Bill of Quantities or BoQ is a tendering document most frequently used in construction and project delivery that presents an itemized list of costs, including materials, parts, and labour.

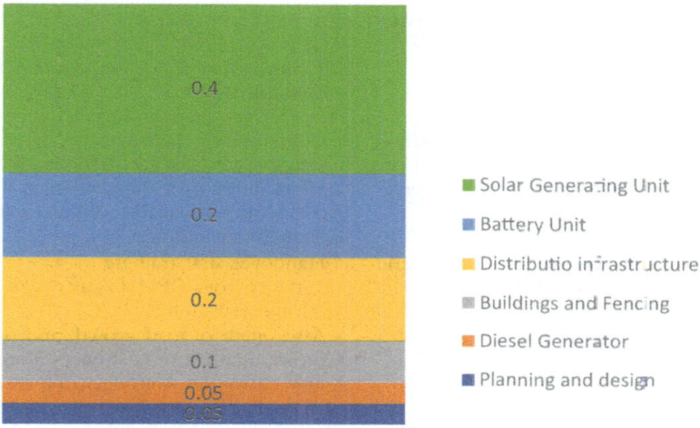

Fig. 5.3 CAPEX for a 30 kW solar mini-grid

additional non-technical CAPEX costs are getting the project up and running, including the cost of obtaining necessary approvals, and licences and navigating regulatory frameworks. Such non-technical considerations are fundamental for the holistic and sustainable development of a CES, aligning technical goals with the broader community context.

A pico-system such as the one proposed in CESET may require further investment down the line to cover additional CAPEX financing needs that may not be covered by the project budget and that will continue after the funded period has ended. These CAPEX costs cover sporadic instances and require investment that cannot be fully accounted for in the current year, all depending on the quality of negotiations with the community (Table 5.1).

High upfront CAPEX can pose a challenge for CES, especially when dealing with individual sites, despite global trends of reduction in costs for components such as photovoltaic (PV) panels and batteries. Implementing bulk purchasing strategies allows for economies of scale, enabling cost efficiencies in acquiring necessary components. Further, improvements in supply chains, marked by reduced transport fees and streamlined logistics, contribute to overall CAPEX reduction. In some countries, policymakers and businesses have explored opportunities to waive import duties and taxes, which can make the deployment of CES more financially feasible and sustainable.

Table 5.1 Additional CAPEX costs unique to CES systems

Types of costs	*Elements that require negotiation with the community*
Consumer/community finance to address the affordability gap of end users and lack of access to financing for small enterprise and community initiatives	Community engagement to determine which appliances are needed, what the affordability is of the customers including payments on loans as well as legal and financial services fees
Handover costs of transferring micro-grid ownership and management to a community body	Handover staff training Support from CESET/current owner during transition
Extension to generation or distribution systems	Assessment of load growth and potential future demand of customers. Potential new customers wanting to be connected

5.3.2 *Operational Expenses (OPEX)*

Unlike CAPEX, which addresses initial capital needs, OPEX caters to the day-to-day expenses incurred during the lifespan of CES. An understanding of the intricacies of ongoing operational costs is key to ensuring CES functionality, optimising resource allocation for routine activities, ensuring the longevity of projects, and crucial for establishing a robust financial framework that contributes to the enduring success and resilience of CES.

Examples of OPEX costs include maintenance contracts, monitoring fees (e.g. data or SaaS), security, fuel, customer service, billing, collection, and land rent. Costs can either be fixed (e.g. the depreciation of assets, interest on debts, fixed taxes, and fees) or vary based on demand or number of customers. A breakdown of routine costs for a 30 kW solar micro-grid is shown in Fig. 5.4, while a summary of different types of OPEX costs is outlined in Table 5.2.

OPEX costs do not typically become known until after financial close, and often after months of steady operations. However, to aid planning, pre-project financial modelling use rules to anticipate what OPEX costs may be. One approach is to estimate OPEX costs as 20% of expected total annual revenues; another is around 5–10% of total CAPEX costs. For example, if CESET has a maximum CAPEX budget of £75,000, OPEX costs could be in the region of £3–10,000 per annum. This value is entirely dependent on the nature of the micro-grid installed, including

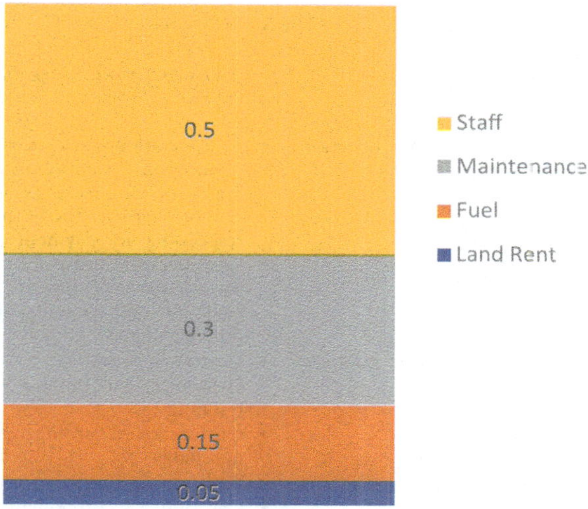

Fig. 5.4 Example OPEX costs for a 30 kW solar-diesel mini-grid

Table 5.2 Standard types of OPEX costs

OPEX type	*Description*	*Example*
Fixed costs	Do not vary with changes in volume demand, peak demand, or no. of customers	Overheads and transactions costs, local operation costs, office costs, management staff
Demand variable costs	Vary with changes in the volume of demand	Fuel costs, lubrication oil, maintenance related to throughput, revenue-dependent taxes
Customer variable costs	Vary with changes in the number of customers	Marketing and outreach campaigns. data fees for smart meters

its scale, the existing and future demand on the grid and the baseline economic situation at the chosen site.

In addition to typical OPEX costs incurred by mini-grid developers, CES may have additional OPEX costs to consider accounting for the enhanced community involvement outlined in Table 5.3.

Table 5.3 Examples of exceptional OPEX costs related to CES

Types of costs	*Elements that require negotiation with the community*
Operational costs involved in the collection and management of metering, billing, and payments	• Consideration must be given to the implications of choosing mobile, token, or cash-based payment systems and of choosing the authority responsible for payment collection • Consideration must also be given to the adequacy of different modes of payment: prepayment, pay-as-you-go, post payment
Operational costs involved in after sales and customer service	• Customer issues • Customer/staff training and capacity building • Customer onboarding, marketing, and upselling
Operational costs involved in technical maintenance	• Call outs for minor repairs • Call outs for component replacements • Call outs for infrastructure repairs, upgrades, and expansion • Safety checks • Further considerations include the need for agreements with third-party suppliers (e.g. metering and payments vendors), and contracts with post-project contractors that will need to be agreed upon to deliver operational support and services
Capacity building	• Investment in community capacity building to enhance financial literacy and understanding of the CES's financial dynamics • Training on safe use of electricity or using electricity to start businesses to promote economic development
Monitoring and evaluation	• Implementing robust monitoring and evaluation mechanisms to track the financial plan's effectiveness, including regular assessments of revenue generation, cost management, and overall financial sustainability • Measuring the social impact of the CES in order to adapt the delivery model to improve value to the community

According to a survey of 13 African Minigrids (International Finance Corporation, 2017), OPEX typically account for 58% of revenue, while when combined with administrative costs, the total expenditure reaches 128% of revenue. Such high OPEX costs are due to high operational expenditure from challenges of reaching remote locations and the need to trial unproven operational strategies, coupled with the fact that revenue is low (IRENA, 2018). The use of smart meters and remote monitoring can reduce OPEX costs by improving maintenance efficiency and reducing staff time. Additionally, CES can engage with the community to carry out routine maintenance to further reduce costs.

5.4 Revenue Model

The financial sustainability for CES, tariff modifications, and business model planning all depend on understanding revenue generation, aiming for a positive balance to be struck between income from electricity sales and operational costs for staff, maintenance, and other running costs. Revenue is earned through connection fees, electricity sales, and grants/subsidies and is reliant on variables including demand for electricity, the ability and willingness to pay and the tariffs set for consumers (USAID[2]). There is, however, an enormous gap in recognising and valorising the multiple benefits provided by community energy beyond producing sales revenues.

Tariffs need to be affordable to customers but also need to be at levels able to generate adequate revenues to meet recurring expenditures and other liabilities and, in some cases, generate an adequate profit and recover the capital cost of the system to be fully commercial (NDC Partnership[3]). Tariffs should be set based on projected demand, and in order for the scheme to be viable, they should cover all the costs, both fixed (e.g. operation, wages) and variable (e.g. maintenance, spare parts, training) of the CES (NREL, 2018). A basic rule generally accepted in rural electrification planning is that, regardless of the scheme chosen, a tariff should at least cover the system's running costs to ensure the ongoing operation of a system through its lifetime.

[2] https://www.usaid.gov/energy/mini-grids/regulation/tariffs, retrieved January 16, 2024.

[3] https://ndcpartnership.org/case-study/smart-incentives-mini-grids-through-retail-tariff-and-subsidy-design, retrieved January 8, 2024.

In crafting a robust tariff model for CES, several crucial factors demand consideration. Operational costs comprising an in-depth analysis of project-related expenses outlined above provide the foundation to determine the minimum revenue required for ensuring the financial sustainability of the project. The technology lifecycle adds an additional layer of complexity. Long-term financial planning into the tariff structure must incorporate the lifespan and depreciation of energy-generation technologies. Additional costs to cover may include interest on loans or equity demands from investors and potential income from subsidies or grants. The total revenue requirement is then compared with community affordability to devise tariffs to cover costs, ensuring tariffs are aligned with the community's ability and willingness to pay. This multifaceted approach ensures the development of a tariff model that is not only financially sustainable but also socially inclusive and considerate of the diverse dynamics within the community.

In delving into the critical aspect of community acceptance, actionable strategies for actively engaging communities throughout the tariff-setting process are required. Prioritising clear and transparent communication becomes paramount, highlighting the costs and benefits intricately linked to the tariff structure to foster community comprehension. Integrating community voices and preferences stands central in the process, employing consultative approaches to gather feedback and align the tariff model with local expectations, fostering a sense of community ownership. The implementation of educational programmes can enhance community understanding by shedding light on the factors influencing tariff rates and emphasising the broader benefits stemming from their contributions.

The ability and willingness to pay varies depending on the geographic location. Areas with larger population densities tend to have more vibrant economies; hence, micro-grids operating in those areas tend to be more profitable than those operating in remote locations (Bhattacharyya, 2018). Ideally, systems designed in rural areas should adopt a pro-poor approach to ensure affordability even for low-income consumers. However, micro- and pico-grids may have different requirements and can organise the tariff system in different ways. In our case, the tariff structure will have to be closely negotiated with the community and a realistic assessment of their capacity to make payments. Examples of tariffs paid by rural mini-grid customers in Africa are outlined in Table 5.4.

Within CES, there exists a spectrum of tariff models, each catering to specific requirements. Key tariff principles for CES include simplicity,

Table 5.4 Examples of cost reflective tariffs on mini-grids in Africa

Country	*CRT*	*System*	*Source*
Tanzania	USD 0.74/kWh	7 kW system	NREL (2018)
Ghana	USD 0.75—0.80 per kWh	100-household village, PV-diesel hybrid	NREL (2018)
Malawi	USD 0.5—1.2 kWh	12 kW system	EASE (2022)

fairness, transparency, justifiability, reasonability, and consideration of seasonality. Figure 5.5 shows some of the considerations involved when choosing different tariffs, while Table 5.5 provides an overview of various tariff types. It is worth noting a pertinent insight from the mini-grid literature, suggesting that pay-as-you-go systems may compromise the operation of mini-grids due to the absence of a consistent revenue stream (Bandi et al., 2022). This underscores the need for thoughtful consideration and adaptation in selecting tariff models to ensure the sustained success and resilience of CESs in dynamic community environments. Cross-subsidisation can be considered, exploring models that allow more affluent users to subsidise access for economically disadvantaged community members, fostering a balanced and equitable energy distribution system.

Navigating the development of a tariff model for CES presents inherent challenges that demand careful consideration. One significant hurdle involves managing the fluctuating energy demand within the community and formulating tariffs that can effectively accommodate these variations. Striking the right balance between simplicity for community comprehension and the necessary complexity to accurately represent the actual cost of energy provision adds another layer of complexity. Additionally, ensuring regulatory compliance is crucial, requiring a delicate approach to uphold established frameworks while also remaining adaptable to meet the unique needs of the community.

5.5 Finding the Finance

Funding is a critical aspect of CES development, influencing their sustainability and impact. From mini-grid experiences, financing the system requires looking beyond the material aspects of the projects. CES will require at least two types of financing:

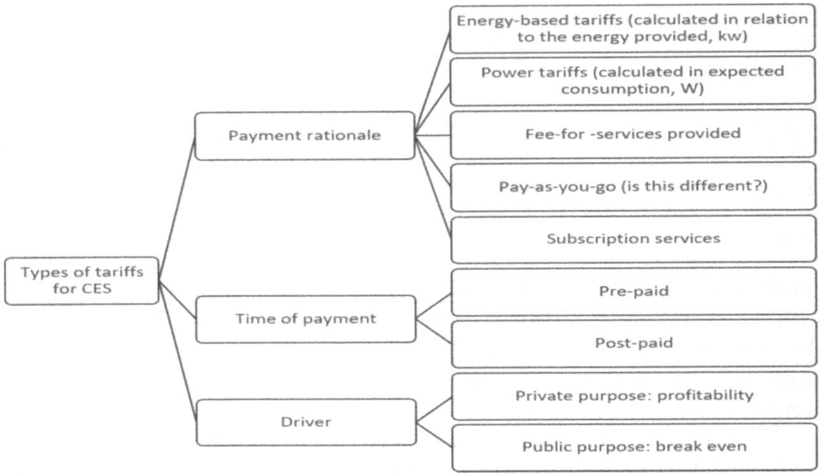

Fig. 5.5 Considerations for tariffs

1. Energy end users, for example, may lack the ability to pay for new appliances or one-time connection fees and, therefore, require financial assistance to be able to receive electricity from the grid.
2. Energy producers, those that install and operate the grid infrastructure.

Table 5.6 outlines the typical financial needs of different stakeholders.

As previously mentioned, the upfront costs of deploying remote infrastructure in rural areas are still high despite recent cost reductions in solar PV and batteries (IRENA, 2015). Additionally, due to uncertain demand, perceived low ability to pay, and challenges relating to maintaining energy infrastructure in remote areas, CES, such as micro-grids, are perceived as high risk by investors and donors (IRENA, 2018). To address these challenges, practitioners and project developers have trialled a variety of financing mechanisms, including public-private partnerships, crowd-funding, and micro-finance programmes. An outline of the key funding models for CES is summarised in Table 5.7.

Public financing, patient, long-term private financing, and public/private financing are all top-down state and market-driven approaches to increasing energy access through the financing of CES. These approaches

Table 5.5 Types of tariffs

Type of Tariff	*Description*
Energy-based	Charges are based on the amount of energy consumed, typically measured in kilowatt-hours (kWh). Users pay for the actual energy they use
Power-based	Charges are based on the maximum power demand or capacity required by the user, often measured in kilowatts (kW). The tariff is determined by the peak power demand
Flat rate or service tariff	Users pay a fixed rate or fee for the service, regardless of the amount of energy consumed or power demanded. It provides simplicity and predictability
Pay in advance or after usage	Payment is made either before using the energy or after consumption. Prepayment models are common in off-grid systems to ensure revenue collection
Limited or unlimited power consumption	Users may have a capped limit on the amount of power they can consume (limited), or they enjoy unrestricted consumption (unlimited) within a specified period

rely on governments or businesses providing capital or subsidies for the construction and operation of energy infrastructure (USAID[4]). The benefits of these approaches include lower upfront costs and access to a larger pool of capital. However, drawbacks can also include a lack of flexibility, high transaction costs, and a lack of local control.

Alternatively, bottom-up, cooperative, and social enterprise models have been deployed to deliver community energy in recent years, providing an alternative to traditional financing models. These models are characterised by increased community involvement and ownership, with local stakeholders actively participating in the design and implementation of CES (Safdar, 2017). While these models require more upfront

[4] https://www.usaid.gov/energy/mini-grids/financing/capital, retrieved January 16, 2024.

Table 5.6 Stakeholder types and financing requirements for these two user types

Stakeholder	*Indicative list of typical financing needs*
Energy End Users • Households • Small enterprises and local livelihoods (including agriculture) • Health, education, and community institutions	• One-time down payment for energy system (e.g. connection fee) • Ongoing payments for energy system (e.g. kWh tariff) • Maintenance fees and service payments • Purchase of efficient appliances/ equipment (particularly small enterprises, local livelihoods and health, education and community institutions) • Upgrading energy system (e.g. higher tariff) • Start-up capital for livelihoods/ enterprises resulting from energy access (productive use of energy)
Energy producers • For-profit enterprises- micro, medium, and small sized • NGOs engaged in service delivery	• Capital for early stage innovation, R&D and installation and procurement • Pilots and demonstrations to prove the service model • Working capital for operations • Consumer finance/credit to address affordability gap of end users • Internal capacity building and training • Credit for growth and expansion • Capital for diversification of products, solution and upgrading technology to meet consumer needs • Credit or fee to enable servicing in distant/ remote areas

effort and resources, they can also provide more autonomy and greater local ownership in the long term. This model points towards the need to examine the costs in practice as they unfold in each context. While micro-grids are dependent on a solid investment plan to attract businesses or other organisations (such as cooperatives) who want to run them, community energy systems may depend on the reduction of operating costs to the minimum.

Table 5.7 Key Funding models for CES

	Advantages	*Disadvantages*
Grants and Subsidies	• Provides upfront capital without the need for immediate repayment • Alleviates financial burden on communities during the initial project stages • Often targeted at renewable energy projects, encouraging sustainable practices	• Dependency on external funding sources, which may be limited or subject to policy changes • Grants may have specific criteria, restricting flexibility in project design
Private Sector Finance (Debt and Equity)	• Can provide significant capital through debt or equity arrangements, supporting the financial needs of CES projects • Attracts investors seeking financial returns, aligning with profit-driven motives and potentially facilitating large-scale funding	• Debt financing requires repayment with interest, increasing financial obligations for the CES project • Equity investors may seek ownership stakes or dividends, potentially compromising community control over the project • Contingent on the financial viability and creditworthiness of the CES project
Public–Private Partnerships (PPPs)	• Combines public and private resources, sharing risks and responsibilities • Attracts private sector expertise and investment	• Complex negotiations and potential conflicts of interest • Profit-driven motives of private partners may clash with community-focused objectives
Community-Based Financing	• Fosters a sense of ownership and empowerment within the community • Aligns with the principles of social and economic development	• Limited capacity in communities to raise substantial capital • Slow pace of fund accumulation, potentially delaying project implementation

(continued)

Table 5.7 (continued)

	Advantages	*Disadvantages*
Impact Investing	• Attracts private capital to match financial return and positive social and environmental impacts • Aligns with the increasing trend of socially responsible investing	• May prioritise financial returns over community needs • Limited availability of impact investors in some regions
Crowdfunding	• Engages a broader audience in supporting community energy projects • Accessible and transparent fundraising model	• Reliance on the community's ability to mobilise support • Uncertainty in achieving large-scale funding goals

Several key considerations play a crucial role when selecting funding models for CES. First and foremost is the level of community involvement and their willingness to contribute financially, emphasising the importance of understanding local dynamics. The scale and complexity of the project are also pivotal factors, with different funding models aligning better with varying project sizes and intricacies. Additionally, stakeholders must evaluate their risk tolerance, accounting for financial stability and uncertainties inherent in the project. Being mindful of the regulatory environment, encompassing local and national regulations governing energy project funding, is essential. Furthermore, the consideration of the long-term sustainability of the chosen funding model extends beyond the project's initial phases, ensuring enduring success and impact.

5.6 Community Energy in Malawi

The Rural Energy Access through Social Enterprise and Decentralisation (EASE) project,[5] whose aim was to progress the SDG7 in Malawi, ran from 2018 to 2024 with funding from the Scottish Government. EASE was coordinated by the University of Strathclyde in partnership with Self Help Africa. The objective was to increase access to sustainable energy

[5] https://ease.eee.strath.ac.uk/.

for rural communities in Dedza and Balaka, enabling economic development and improved livelihoods. Two solar micro-grids were installed in the Dedza district through EASE, generating and distributing power for localised domestic and productive uses (Fig. 5.6).

The key lessons learned from these installations were:

- Capital and operational costs were high when compared with established benchmarks, underscoring the emergent nature of this market in Malawi.
- Demand and ability to pay for electricity services were both found to be high, despite the rural location and low incomes of the community.
- While revenue generated from electricity sales adequately covers on-site operational expenses encompassing maintenance contracts, data management, and site agents, it falls short of covering broader organisational costs like transportation and staff salaries.

Fig. 5.6 12 kW solar micro-grid, Mthembanji, Malawi

Table 5.8 Technical overview of Malawi micro-grids

	Kudembe	*Mthembanji*
Installed	2022	2020
Number of customers	50	60
PV Generation	11 kW	12 kW
Distribution grid	240 V single phase	
Battery Capacity	20 kWh Li-ion	19.8 kWh Li-ion
Inverter manufacturer	SMA	
Installation and Maintenance	BNG Electrical, Lilongwe	
Smart Meters	Steamaco	

- Community micro-grids in Malawi depend on continued donor support and subsidies to achieve financial sustainability.

A summary of the technical parameters comprising solar PV, lithium-ion batteries and a single-phase distribution grad are summarised in Table 5.8.

5.6.1 *Capital and Operational Costs*

Capital costs have been found to be high. These are pioneering projects in Malawi and costs are expected to reduce as more micro-grids are installed creating economies of scale. Transport costs from South Africa increased the costs further.

In the future, the strengthening of local supply chains for solar equipment may drive these costs, including transport, down. However, high costs are also due to inflation and foreign exchange rate fluctuations, pushing local fuel and labour prices up and resulting in significant cost increases for local component and contractor costs. Macroeconomic volatility has a direct impact on local supply chains and micro-grid project costs and is likely to be a key influencing factor on future micro-grid CAPEX in Malawi (Fig. 5.7).

Site-based operational costs for one of the micro-grids total USD 316.4 per month on average or USD 3,796.80 annually. Operational costs include site agent and security guard salaries, data and SaaS fees for smart meters, and a generation and distribution maintenance contract, but do not include field and management staff costs, transport costs

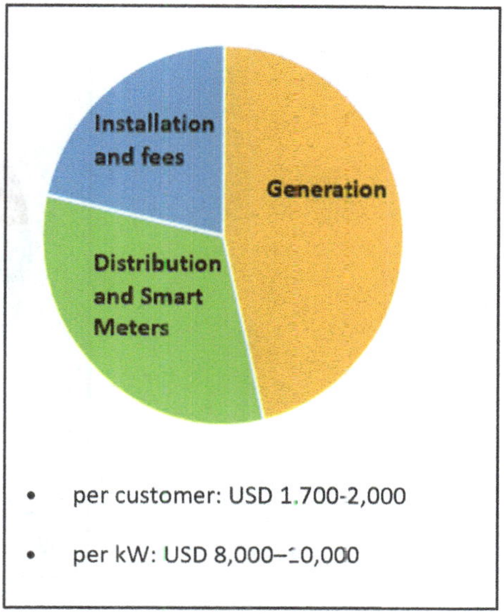

Fig. 5.7 Malawi case study CAPEX

and business overheads, as these have been covered through EASE grant funding.

The cost per customer per month (USD 4.27) is on the high side of benchmark estimates for sub-Saharan Africa, which tend to be in the bracket of USD 2.50–6.00 (AMDA, 2020) (Fig. 5.8). A comparison with monthly revenue reveals income only just covering site-based costs, compromising financial sustainability without interventions on tariffs or demand.

The majority of OPEX costs come from a maintenance contract with a Lilongwe-based electrical contractor. This is currently the only option given the lack of technical capacity to conduct robust maintenance on micro-grids. There is great potential to reduce these costs through in-house maintenance technicians, with salaries paid through central funds, and only paying for transport/material costs needed for maintenance trips. Travel to different micro-grid sites could be combined, and efficient logistics strategies could be employed to reduce travel times and save on costs.

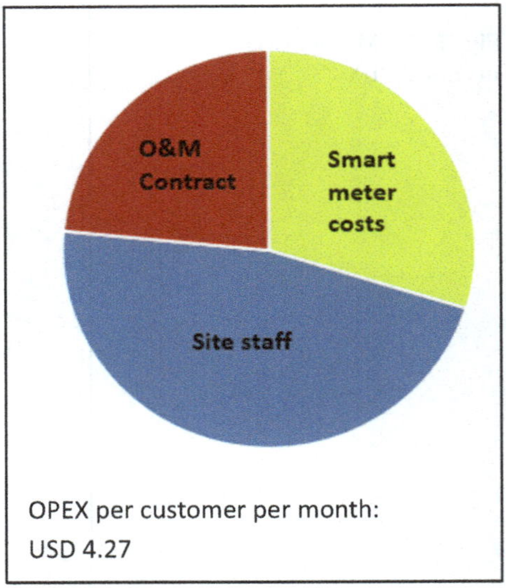

Fig. 5.8 Malawi case study OPEX costs

5.6.2 *Setting Up Tariffs*

Tariffs are paid through site agents in a PAYGO format, where customer balances are topped up through the SteamaCo platform.[6] The tariffs have been set and adjusted through ongoing community engagement and negotiations on willingness to pay, with different tariffs designed to cater for different customer segments, as outlined in Table 5.9.

The Banja tariffs offer a set allocation of energy for a daily fee, which allows for domestic use including lights, phone charging, and TV. The Ufulu tariff for business customers is tiered and reduced for higher energy users. A significant daytime discount (75%) promotes demand when excess electricity is available during sunlight hours.

Figure 5.9 shows Average Revenue Per User (ARPU) per month for 2021, disaggregated by customer segment. Residential ARPU follows a seasonal trend, with higher spending corresponding to the rice harvest season in July, while business ARPU is considerably higher and follows a less prominent seasonal trend. The mean ARPU for the year is 5.43

[6] More information on the commercial page. https://steama.co/#home, retrieved January 16, 2024.

Table 5.9 Tariff summary for Mthembanji and Kudembe

Bundle	*Services*	*Payment type*
Banja Monthly (Household)	A set allocation of energy (260 Wh per day) which approximately equates to a daily service of: – 3 lights for 3 hours – 1 light for 8 hours – 1 hour of TV – 2 hours of phone charging	Monthly service fee
Ufulu (Freedom)	Unlimited electricity paid for per unit. A cheaper rate applies for higher use	Pay as you Go
Ufulu Daytime	Daytime discount 75% reduction in standard Ufulu costs	Pay as you Go
Midzi (Community)	Electricity for Schools, Churches, or other community groups based on your needs	Pay as you Go

USD/month, which is higher than estimates for Tanzania ($4.58), Kenya ($2.96), and Nigeria ($4.83) (AMDA, 2020).

The seasonal trends corresponding to harvest seasons can be used to plan timings of appliance financing programmes or seasonal tariffs. Acknowledging the mean ARPU of businesses (USD 8.48) is more than double residential (USD 3.89) highlights the importance of increasing revenue through promoting productive Uses of Energy with targeted

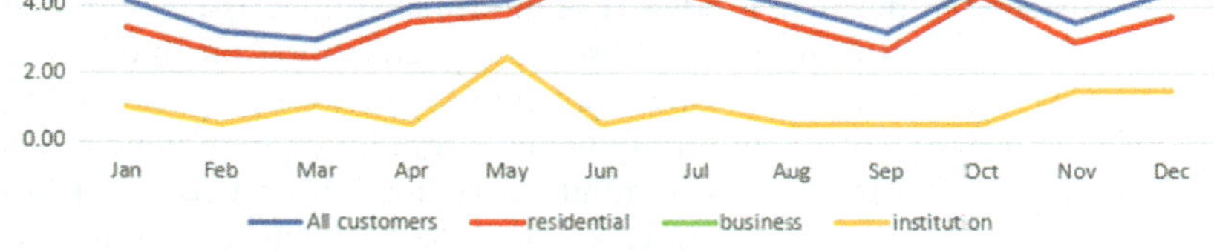

Fig. 5.9 Customer disaggregated Average Revenue per user per month (USD), 2021, Mthembanji

business support. In the case of Mthembanji, the income only covers the monthly OPEX costs, and provides no support for additional staff costs, transport, or wider business costs.

The ARPU data provides valuable insight into rural customers' ability and willingness to pay. The community initially found the tariff too high, resulting in complaints and negotiations conducted over time to find an acceptable tariff. Ongoing assessment of willingness to pay is essential for finding appropriate tariffs, ensuring customer satisfaction and sustainable electricity consumption levels that don't further impoverish communities. Data sharing of ARPU between micro-grid developers progresses the knowledge base to inform sustainable business models with affordable tariffs.

5.7 Conclusions

This chapter has outlined the basic financial features of CES development. Selecting appropriate financial management approaches and funding models are strategic decisions that necessitate careful consideration of community dynamics, project characteristics, and the broader socio-economic context. By exploring and understanding the advantages and disadvantages of different approaches, CES developers can tailor their strategies to ensure both short-term success and enduring impact.

Advancing the understanding of the financial management of CES requires closer collaboration between academic institutions and practitioners, leveraging data analysis and knowledge exchange. Techno-economic business modelling should be a priority, with a focus on developing and testing CES business models linked to innovative financing mechanisms. Additionally, research should emphasise CES performance monitoring through data acquisition and analysis, understanding demand patterns, and exploring productive use opportunities and their contribution to sustainable business models. Longitudinal studies assessing social impact, conducted through cross-disciplinary collaboration and social impact surveys following established frameworks and best practice guides, will provide insights into community benefits, guiding recommendations for interventions to increase community participation and impact from electricity connections. These research areas collectively contribute to accelerating CES deployment and ensuring financial sustainability.

The introduction of smart subsidies is essential to address the financial challenges faced by CES. These subsidies, supported by the government, can enable CES to connect and provide reliable electricity services to rural communities, balancing affordable tariffs with operational sustainability. A well-designed subsidy system, based on data sharing among active CES projects, can be economically modelled to determine the necessary support. Removing barriers such as VAT and Import Tax on CES components can significantly reduce capital expenditure, fostering a more favourable financial environment. Investing in research and capacity building is crucial, involving efforts to develop skilled technicians, system designers, and business expertise through government-supported training programmes, business development initiatives, and collaboration with academia on research and development initiatives. Collectively, these policy recommendations aim to create an enabling environment for the sustainable financial management and deployment of CES.

The main financial question is whether a CES can be integrated within a community in a way that the community can reduce its operating costs and support its long-term viability. It follows that a process of negotiation of community governance may help redefine the terms of implementation and, hence, support the viability of alternative finance models or subscriptions. The question that follows is which of those costs could be supported by the community. These are two complicated questions which we hope we will be able to answer within the life of the project CESET.

Acknowledgements This chapter is inspired by and expands preliminary work presented in CESET's briefing note "Financial aspects of micro- and pico-community energy systems", available at https://cesetproject.com/sites/default/files/Financial%20aspects%20of%20community%20energy%20systems.pdf.

References

AMDA African Minigrids Development Association. (2020). *Benchmarking Africa's Mini-grids*. Retrieved January 8, 2024, from https://shellfoundation.org/learning/amda-benchmarking-africas-mini-grids/

Bandi, V., Sahrakorpi, T., Paatero, J., & Lahdelma, R. (2022). The paradox of mini-grid business models: A conflict between business viability and customer affordability in rural India. *Energy Research & Social Science, 89*, 102535. https://doi.org/10.1016/j.erss.2022.102535

Bhattacharyya, S. C. (2018). Mini-grids for the base of the pyramid market: A critical review. *Energies, 11*(4), 813. https://doi.org/10.3390/en11040813

Dauenhauer, P., Frame, D., Eales, A., Strachan, S., Galloway, S., & Buckland, H. (2019). Sustainability evaluation of community-based, solar photovoltaic projects in Malawi. *Energy Sustainability and Society, 10*(1), 1–20.

EASE. (2022). *Deploying Solar Microgrids in Malawi.* Technical Report. Retrieved January 16, 2024, from https://ease.eee.strath.ac.uk/wp-content/uploads/2022/11/Deploying-Solar-Microgrids-in-Malawi-Technical-Report.pdf

International Finance Corporation. (2017). *Operational and financial performance of mini-grid descos.* Retrieved January 8, 2024, from http://sun-connect-news.org/fileadmin/DATEIEN/Dateien/New/IFC_Minigrids_Benchmarking_Report_Single_Pages_January_2017.pdf

IRENA. (2015). *Off-grid renewable energy systems: status and methodological issues.* Working Paper. Retrieved January 16, 2024, from https://www.irena.org/-/media/Files/IRENA/Agency/Publication/2015/IRENA_Off-grid_Renewable_Systems_WP_2015.pdf

IRENA. (2018). *Policies and regulations for renewable energy mini-grids.* Retrieved January 16, 2024, from https://www.irena.org/-/media/Files/IRENA/Agency/Publication/2018/Oct/IRENA_mini-grid_policies_2018.pdf

NREL. (2018). *Tariff considerations for microgrids in sub-Saharan Africa.* Retrieved January 8, 2024, from https://www.nrel.gov/docs/fy18osti/69044.pdf

Safdar, T. (2017). *Business models for mini-grids.* Smart Villages Technical Report 9. Retrieved January 8, 2024, from http://e4sv.org/wp-content/uploads/2017/05/TR9.pdf

Weston, P., Kalhoro, W., Lockhart, E., Reber, T. J., & Booth, S. S. (2018) *Financial and operational bundling strategies for sustainable micro-grid business models* (No. NREL/TP-7A40-72088). National Renewable Energy Lab.

CHAPTER 6

Regulating Community Energy at the National Level Comparing Ethiopia, Malawi and Mozambique

Carlos Shenga, Lorraine Howe, Idalina Baptista, Getachew Bekele, and Maxon L. Chitawo

Abstract There is a generalised assumption that the development of energy policy and regulation leads to increased access to energy. This paper investigates the empirical evidence to support this assumption in Ethiopia, Malawi and Mozambique, providing a comparative assessment of the regulatory landscape of energy in the three countries and their

C. Shenga (✉) · L. Howe
Centre for Research on Governance and Development, Maputo, Mozambique
e-mail: cshenga@cpgd.org.mz

L. Howe
e-mail: lhowe@cpgd.org.mz

I. Baptista
University of Oxford, Oxford, UK
e-mail: idalina.baptista@conted.ox.ac.uk

G. Bekele
Addis Ababa University, Addis Ababa, Ethiopia
e-mail: getachew.bekele@aait.edu.et

V. Castán Broto (ed.), *Community Energy and Sustainable Energy Transitions*, https://doi.org/10.1007/978-3-031-57938-7_6

current status in terms of advancing towards universal energy access for both electricity and fuels. Using comparatively available data, the analysis examined the impact of extensive and dispersed bodies of regulation on energy access, as well as the extent to which pioneering community energy enhances energy access. The results are examined in the light of the current context of energy provision in the three countries. The results suggest that universal access to electricity requires an extensive body of energy policy in general and regulation of community energy in particular. However, while being a pioneer in community energy is correlated with improvements in energy access, the factors that explain such a correlation are not clear. More research needs to advance the current understanding of how regulation interacts with other drivers of infrastructure development and innovation to understand what works in a sustainable transition to provide universal access to clean energy.

Keywords Regulation · Energy access · Extensive body of legislation · Policy pioneers · Ethiopia · Malawi · Mozambique

6.1 Introduction

Does regulating community energy facilitate access to electricity? There is a general assumption that the adoption of energy policy and regulation enhances access to energy (Baumli & Jamasb, 2020; ESMAP, 2022; Huhta, 2022; Mahmood et al., 2021; To, 2015). For example, clear energy policy and regulation help to attract private investment for energy development (Howe & Shenga, 2023) as financial institutions invest on the basis of the effectiveness of the policy and regulatory framework (Baumli & Jamasb, 2020; ESMAP, 2022). This includes energy policies beyond those that specifically support access to electricity because access to electricity occurs within a broader regulatory framework. In this chapter, we analyse the extent to which having a developed regulatory framework for energy, as a whole, influences energy access, directly

M. L. Chitawo
Mzuzu University, Mzuzu, Malawi
e-mail: chitawo.m@mzuni.ac.mw

or indirectly and the extent to which pioneering policies for community energy increases access prospects.

To develop this analysis, we focus on a comparative analysis of the impact of regulations on energy access in Ethiopia, Malawi and Mozambique.The three countries belong to the Eastern Africa region, although Mozambique and Malawi are also counted geopolitically as Southern Africa countries and are members of the Southern Africa Development Community (SADC). Ethiopia has never been colonised and hosts the headquarters of the African Union. Malawi was colonised by Britain and Mozambique by Portugal, but today both belong to the Commonwealth. Mozambique joined voluntarily in 1995 (Table 6.1). All three are classified as low-income developing countries[1] and are located in the third quartile of the Human Freedom Index.[2] Ethiopia is a federal state, classified in 2022 by Freedom House as a "not free" country. Malawi and Mozambique are unitary states and rated "partly free".[3] All transitioned to a multiparty system in the early 1990s. In Ethiopia, political distress led to a violent conflict in the region of Tigray in 2021. In Malawi, multiparty elections have produced alternations of political power. In Mozambique, the revolutionary party Frelimo has ruled the country since independence in 1975. Ethiopia is the second most populated country in the continent with 115 million inhabitants while Mozambique's population is 31 million and Malawi's 19 million.[4]

6.2 Energy Policy and Access to Energy

The energy literature suggests that access to energy is enhanced when countries and their key actors advance energy policies and regulations (Huhta, 2022; To, 2015). Regulatory quality seems to correlate positively with increases in renewable and nonrenewable energy consumption

[1] As classified by the World Bank and UN. https://data.worldbank.org/income-level/low-income, retrieved January 23, 2023.

[2] The Human Freedom Index comprises measures of human freedom, personal freedom and economic freedom. https://worldpopulationreview.com/country-rankings/freedom-index-by-country, retrieved March 30, 2023.

[3] 2022 Freedom in the World. https://freedomhouse.org, retrieved March 30, 2023.

[4] Ethiopia is the second most populated in the continent after Nigeria. https://www.worldometers.info/population/countries-in-africa-by-population/, retrieved March 29, 2023.

Table 6.1 Key features of the compared countries

Key features	*Ethiopia*	*Malawi*	*Mozambique*
Population (in 2023)	114,963,588	19,129,952	31,255,435
Politics	Political distress led to Tigray conflict in 2021	Multiparty elections led to alternations of power	Multiparty elections cemented the dominant party system
Past before independence	Never been colonised	Colonised by Britain and became a member of the Commonwealth	Colonised by Portugal and voluntarily joined the Commonwealth
System of government	Federal state	Unitary state	Unitary state
Freedom status	Not free	Partly free	Partly free

Source The authors—compiled from different sources

(Mahmood et al., 2021). This may bode well with ongoing efforts by many countries to embed new policies to improve their energy independence (ESMAP, 2022). Indeed, more than half of the world's population without access to energy live in countries with only low or intermediate levels of energy policy and regulation, while some 400 million people without access live in countries with no energy legislation or regulation (ESMAP, 2022). Moreover, the role of energy policy is crucial to manage new challenges, with increased demand for regulation and fiscal incentives on imports of energy technology (e.g. tax reductions or exemptions on solar systems) or to access large-scale energy investment (ESMAP, 2022). The primary concern for investors is whether or not they trust the effectiveness of the energy regulatory framework (Baumli & Jamasb, 2020). Adequate regulations and incentives like tax reductions may foster an enabling environment that attracts private investment (ESMAP, 2022).

While energy policy plays a fundamental role in energy access, not all energy policies are the same. In the hierarchy of laws, policies include all legislation from the first-tier legislation to lower (second, third, fourth, etc.) tiers. Included would be legislation from the constitution of a given country (first-tier) to its laws or acts, proclamations, decree-laws, decrees, ministerial diplomas, resolutions, directives and so on.[5] Regulating community energy constitutes only one part that addresses energy policy or legislation. The other is the overall body of energy legislation and regulation. Our concern with the effect of energy policy is twofold: to test the effect of ***the body of energy legislation and regulation*** in general and of ***community energy regulation*** in particular. The former consists of general energy legislation and regulation of a given country. The latter refers to energy regulation that focuses primarily on small-scale off-grid energy systems, mostly in rural and hard-to-reach communities.

6.3 Hypotheses

We pose three hypotheses for the relationship between energy policy and access to electricity. The first two focus on the effect of the body of energy legislation and regulation, and the third on community energy regulation.

> *H1:* Having an extensive body of energy legislation and regulation increases access to electricity more than having less extensive legislation and regulation.

The assumption underlying this hypothesis is that: on the one hand, having an extensive body of energy legislation and regulation indicates over a long period of time that more policymakers are reflecting and choosing better solutions to boost access to energy and more implementers are implementing those solutions. On the other hand, a less extensive body of energy legislation and regulation suggests poor energy accomplishment.

> *H2*: Having a dispersed body of energy legislation and regulation boosts access to electricity more than having a concentrated body of legislation and regulation.

A concentration of energy policies in a few points of time places huge pressure on energy operators to understand and interpret policies quickly, which can, in turn, threaten the energy supply, especially in low-capacity

[5] Due to country differences on terminology, this list does not follow (vertical) hierarchy of law.

contexts like those that characterise many countries in Africa. Alternatively, a dispersed body of energy legislation and regulation across many points of time provides energy operators with more time to learn and adapt to the body of legislation and regulation before implementation.

> *H3*: Being a pioneer[6] in regulating community energy increases access to electricity more than being a latecomer in regulating community energy.

On the one hand, being a pioneer in regulating community energy suggests having more time to build experience, knowledge and skills to master the delivery of electricity by and for communities. It also suggests that the regulation was approved a long time ago, and it has allowed other relevant regulations to come into effect to consolidate it. On the other hand, regulating community energy more recently indicates there is less experience in expanding access to electricity through off-grid solutions to and by communities that have been deprived of the grid system.

6.4 Country Contexts of Energy Sector Development

Before testing and examining these hypotheses quantitatively, it is important to situate the dynamics of energy access in the context of each specific country.

Ethiopia. Electricity was introduced in Ethiopia during the reign of Menilik II around 1898. Electric power supply to some towns, such as Nazareth, Dire Dawa and Dessie, began around 1929. However, in 2000, only 13% of the population had electricity access. Lack of progress in access to electricity was the norm until the late 2010s. After the successful implementation of the First National Electrification Program (NEP) in 2017, significant progress has been made in which 33% of the population was connected to the grid and 11% to off-grid pre-electrification, totalling 44% access (NEP 2.0). There has also been progress in engaging the private sector through policy. The model of Public–Private Partnerships (PPPs) has been advanced to transform energy projects' financing and implementation. The government expects that PPPs will support the

[6] Those that have adopted regulation that recognizes and facilitates community energy earlier than others.

future development of geothermal, hydro, solar and wind power generation projects by giving access to the energy sector to Independent Power Producers (IPPs) and using different project finance modalities, so that the public sector concentrates on regulatory and off-taker roles.

The plans stated in NEP 2.0 to achieve its stated goal of *universal electricity access nationwide, are to provide* 65% with grid solutions, and 35% with off-grid technologies (solar off-grid and mini-grids). The Ethiopian Electric Utility (EEU) is responsible for the implementation of on-grid projects. By 2025, EEU will connect over 15 million households by implementing a top-down approach, compared to 6.9 million in 2017 (reaching the 65% goal). Off-grid projects require a combination of public and private efforts. A complementary bottom-up approach would also be carried out alongside the top-down approach to connect the remaining 6 million rural and deep rural households to reach the 35% set goal. As of 2021, NEP 2.0 has increased the number of connected households to 51%.

Another plan for achieving 100% access by 2025 focused on *"cross-sectoral linkages with the productive and social services sectors and in support of vulnerable groups"*. The focus of this plan is social and economic infrastructures—such as primary and secondary schools, hospitals, primary health centres and productive facilities. Such collective infrastructures have priority in facilitating connections (grid or off-grid) with high economic growth potential, particularly in the agriculture sector.

Malawi. Access to electricity in Malawi is very low, estimated at 18% (12% on the national grid and 6% off-grid systems). It is the lowest rate in the SADC region and among the three countries studied in this chapter. The disparity in electrification rates between the urban and rural strata of the population is significant, estimated at 49% and 4%, respectively. The low rural electrification rate means that community energy systems in Malawi have a crucial role to play in enabling electricity access in rural areas.

The market structure of the electricity sector in Malawi is governed by energy laws enacted in 2004. The enactment of the energy law or legislation provided a legal mandate for restructuring the sector and establishing an independent energy regulator.

Since 2004, there have been several reforms that have been implemented in the sector. One of the most significant was the unbundling of the Electricity Supply Corporation of Malawi (ESCOM), a state-owned electricity utility company, in 2016 into state-owned companies mandated to carry out specific functions in generation, transmission and distribution, Single Buyer (SB) and System Market Operator (SMO). As a result of these reforms, the electricity market in Malawi consists of several players that generate electricity at both large and small scales. Large-scale electricity generators can broadly be classified into two main groups: Major Activity Producers (MAP) and Auto Producers (AP). MAPs are companies whose main activity is to generate electricity. These include the state-owned Electricity Generation Company (Malawi) Limited (EGENCO), which was formed after the unbundling of ESCOM and IPPs. IPPs are private utility companies which own facilities to generate electric energy for sale into the national grid to the public utility and end users. APs generate electricity largely for their own use, but their main economic activity is not electricity generation. Examples include sugar mills.

At medium and small-scale levels, players in the electricity sector in Malawi can be categorised into those that generate electricity in stand-alone systems for their own use and developers of mini-grids that generate and supply electricity to community households. These community energy systems are categorised as mini-grids. A mini-grid framework was developed in 2017 to specifically guide developers of these systems in accordance with Malawi's energy legislation.

Mozambique. The development of Mozambique's electricity sector can be divided into at least three phases. The first, from independence in 1975 to c. 1994, is a phase of Formation characterised by the vertical integration into a single company, EDM, created in 1977, of the dispersed electricity infrastructure built during the colonial period.[7] During this phase, any energy policy was incipient. There were a few political pronouncements about the role energy would play in supporting the overall development strategy, but these were not followed by substantive legislation (Frelimo, 1977). Interventions towards electricity access were largely piecemeal, addressing localised needs in generation and distribution, especially in urban areas (Dava & Tamele, 2011). The decade between 1984 and 1994 was afflicted by substantive difficulties caused by the ongoing civil war and economic crisis. The structural adjustment programme initiated in 1987 anticipated the restructuring of the electricity sector that was to come. The end of the civil war in 1992, followed

[7] Decree-law 38/77.

by multiparty elections in 1994, marked the end of this first phase and augured a shift in policy for the sector.

The second phase, from 1995 to 2014, is Modernisation. During this period, policy slowly moved to adopt the mainstream electricity sector approaches espoused by global entities like the World Bank. This included legislation to facilitate the unbundling of the sector, fostering privatisation and competition, alongside the creation of a regulatory agency.[8] Initial efforts at promoting competition were not very successful, and they did not seem to create significant changes in the drive for electrification. Similarly, efforts to promote investments in the renewables sector via a national renewable energy fund (FUNAE), which was created in 1997,[9] did not significantly change the pace of electrification. Until the mid-2000s, access to electricity remained at around 5–6%. A new impetus towards electrification came during the Presidency of Armando Guebuza, from 2005 to 2014. His government introduced a new electricity strategy that sought the electrification of all headquarters of district units.[10] It also implemented new legislation to facilitate large-scale infrastructure projects, including those led by IPPs.[11] Finally, the Mozambican ownership of the Cahora-Bassa hydroelectric dam (HCB) also allowed for an increase in the share of electricity generated by HCB to be made available to supply the growing internal demand. As a result, access to electricity increased to 26% by 2014 (EDM, 2017).

The third phase, in trend since 2015, is Energy for All. This phase continues to be underpinned by the logic of liberalisation of the previous period, alongside a push towards universal electricity access by 2030, in response to global sustainability agendas.[12] This tension between electricity as a commodity and electricity as a common good has been addressed by new policy strategies and legislation that attribute differentiated roles for the government and the private sector, with an expectation that the donor community will play a key role in providing access to

[8] Decree 28/95.

[9] Decree 24/97.

[10] Resolution 10/2009 and Resolution 62/2009.

[11] Law 15/2011.

[12] Law 11/2017 and Resolution 48/2018.

electricity in areas where profit levels are low.[13] A new off-grid regulation adopted in December 2021 is expected to transform electricity generation, facilitating the entry of Independent Power Producers.

6.5 Data and Research Design

Considering the different conditions for the governance of energy in Ethiopia, Malawi and Mozambique, how does improvement in regulation impact electricity access? To test the hypothesis proposed above, the analysis uses a consolidated set of data compiled by the authors from multiple sources (see below). Data on access to electricity comprises 21 points in time from 2000 to 2020 for the three countries compared. Data on energy policy spans from the period that a given country attained sovereignty to 2022. For Malawi and Mozambique, the data comprises energy policies from the period after they became independent, respectively, in 1961 and 1975. For Ethiopia, it starts in 1955—the year when the first energy legislation was approved through the establishment of the utility company. Additional inferences on governance are made with reference to the analysis of the landscape of energy governance in each country, as summarised above.

The analysis deploys a comparative technique called Most Similar System Design (MSSD), which "*seeks to compare* [cases] *that share a host of common features in an effort to neutralize some differences while highlighting others*" (Landman, 2009: p. 70, Landman & Carvalho, 2017). MSSD seeks to identify the key features that are different among similar countries and which account for the observed outcome" (Landman 2009, Landman & Carvalho, 2017). According to Przerworski and Teune (1970), MSSD is particularly suited for those engaged in area studies—countries that share geographical areas, such as Asia, Africa, Europe and Latin America, regardless of having different histories, languages, religions, politics or cultures. In this case, the three countries share the Eastern Africa geographical region.

[13] Resolution 49/2018 and Law 12/2022.

6.5.1 Dependent Variable—Access to Electricity

Access to electricity is the outcome to be explained—the dependent variable (DV). It is measured by the percentage of the population with electricity to power a basic bundle of energy services, using data from the World Bank Global Electrification Database "Tracking SDG 7: The Energy Progress Report" (WB SDG7).[14] Other studies have measured access to electricity using night light data calculated using satellite images of the earth (Kroth et al., 2016; Min, 2015), household surveys (World Bank, 2015) or community-level data from Afrobarometer based on observations.[15] The first two datasets measure access to electricity using objectively verifiable data as it is taken through observations. The latter measures it subjectively using people's opinions and based on their own experiences.

The focus here is on the WB SDG7 data, complemented with additional sources. The WB SDG7 data is made up of counts calculated by utility companies from each connection they make. The World Bank's custodian agencies obtain these data from governments and assemble it in their databases. However, this data tends to lack uniformity. Moreover, like any data which is submitted within political institutions, there are questions about the possible manipulation of the data. Nevertheless, the data provides a useful comparative perspective on the dependent variable.

The data in Fig. 6.1 shows that access to electricity is higher in Ethiopia and lower in Malawi and that is increasing significantly faster in Ethiopia and Mozambique than in Malawi. Mozambique and Malawi had about the same level of access to electricity in 2000, but Mozambique (12%) started significantly distancing itself from Malawi (7%) by 2005, following Ethiopia.

[14] See https://openknowledge.worldbank.org/handle/10986/33822, retrieved May 10, 2022.

[15] The Afrobarometer instructs interviewers to collect electricity data through observation: whether in the Primary Sampling Unit, there is an "electricity grid that most houses could access". www.afrobarometer.org, retrieved November 25, 2022.

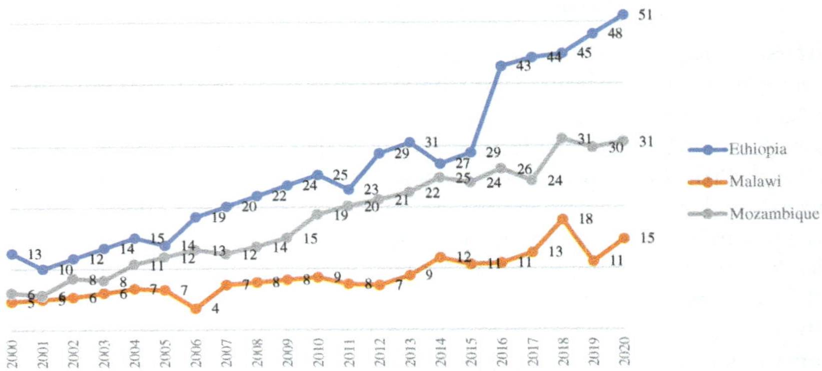

Fig. 6.1 Access to electricity by country and over time, 2000–2020 (percentage) (*Source* World Bank Global Electrification Database from "Tracking SDG 7: The Energy Progress Report" led jointly by the custodian agencies: The International Energy Agency [IEA], the International Renewable Energy Agency [IRENA], the United Nations Statistics Division [UNSD], the World Bank and the World Health Organization [WHO])

6.5.2 *Independent Variables—Regulating Community Energy and the Body of Energy Legislation and Regulation*

Energy policy is addressed by *the body of energy legislation and regulation* in general and *community energy regulation*. Its data is objectively verifiable from official government sources (government gazettes and publications), where it was gathered by the authors. Figure 6.2 shows that Mozambique has the most extensive body of energy legislation and regulation, with 32 items, while Ethiopia has moderately extensive (23 items) and Malawi is less extensive (17 items).

Malawi's body of energy legislation and regulation spans a shorter period from 2003 to 2021, indicating a more concentrated body of energy policy. With respect to Ethiopia and Mozambique, removing outliers, it can be said that Mozambique has a more dispersed body of energy policy while Ethiopia has a concentrated body in two periods: 1996–1999 and 2013–2022. Ethiopia established the utility company (Ethiopia Energy Light and Power Authority, EELPA) in 1956, while for Mozambique, it was in 1977 that the country established its utility

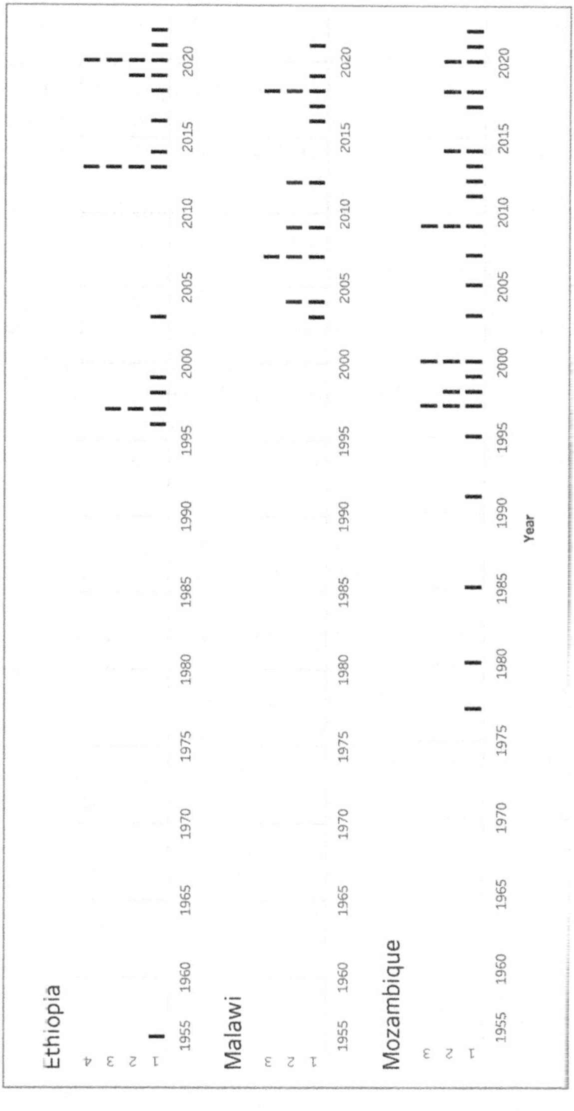

Fig. 6.2 Energy policies by country and over time, 1956–2022 (Generated by the authors from energy policies collected from the respective countries and their own expertise)

company (EDM), after which it experienced a 16-year-long civil war (1977–1992).

Moving to community energy regulation, Ethiopia is the earlier adopter, while Malawi and Mozambique commenced regulating community energy more recently. Ethiopia started regulating community energy two decades ago and continued to do so in the following years. In 1997, the Government of Ethiopia opened up to domestic investors the possibility to invest in plant capacities of up to 25 MW.[16] In the following year, it introduced attractive package incentives and tax reliefs for private investment.[17] In 2020, it issued licenses for electricity supply in the national grid, regulated investments in any sector, including energy, and set out the areas in which foreign investors could invest.[18] In 2019, the Government of Malawi regulated mini-grid energy systems (MERA, 2020).[19] Mozambique lags behind as it only approved its regulatory framework for off-grid energy systems in December 2021, when it established the norms and principles for communities to invest in off-grid energy systems through mini-grids up to 10 MW.[20] The Government of Mozambique has yet to publish all of the related accessory instruments (Howe & Shenga, 2023[21]).

These findings are similar to those from ESMAP's Regulatory Indicators for Sustainable Energy (RISE), where Ethiopia ranks ahead on frameworks for mini-grids (with a score of 73), followed by Malawi (66) and Mozambique (61) (ESMAP, 2022). On frameworks for off-grid systems, Mozambique ranks ahead (with a score of 95) but then follows Ethiopia (83) and Malawi (48) (ESMAP, 2022). This advance of Mozambique over Ethiopia and Malawi may be explained by the role played by Mozambique's National Fund for Energy (FUNAE),[22] a state agency established in 1997 to fund and implement energy projects funded by donors to increase rural access to electricity.

[16] Proclamation 37/1997.

[17] Regulation 36/1998.

[18] Regulation 474/2020.

[19] Government of Malawi, 2019.

[20] Decree 93/2021.

[21] Since publication, 5 of the 27 outstanding accessory instruments were published in the Government Gazette in May 2023.

[22] Decree 24/97.

Based on the different data sources referred to above, the authors consolidated an original dataset on Ethiopia, Malawi and Mozambique's energy policies, access to electricity and other related aspects covering 22 years from 2000 to 2021 and including 66 observations (*n*). Due to data availability for certain years in each country, the number of observations is lower on certain variables.

6.5.3 Model

To test the effect of regulating community energy on access to electricity, this study uses an ordinal logistic regression model since our dependent variable (DV)—*access to electricity* is measured at the ordinal level with a three-point scale: high access (coded 2), medium access (1) and low access (0). One or more independent variables are ordinal or categorical and dichotomous. *Regulating community energy* is a dichotomous variable categorised by advanced (coded 1) and less advanced (0). *Body of energy legislation and regulation* is categorised first by extensive (2), moderately extensive (1) and less extensive (0); and second by dispersed (2), concentrated (1) and highly concentrated (0).

6.6 Empirical Results

Table 6.2 reports the ordinal multiple regression model that tests the effect of energy policy on access to electricity. The results show, overall, that the model is fit since it is significant, it reveals goodness-of-fit as its significance levels are greater than 0.05, and it meets the proportional odds assumption as the test of parallel lines reveals significance greater than 0.05 (Table 6.2). The verification criteria to un/confirm the hypotheses are: first, the level of significance (Sig.), which has to be less than 0.05 for the independent variable (IV) to have an effect on the dependent variable (DV). Second, the magnitude (high/low) and direction (−/+) of IV (seen from B coefficient). The former tells us which IV has high or strong explanatory power on DV and the latter how the IV affects the DV.[23]

The results reveal more specifically that, of the three IVs addressing energy policy, two are statistically significant (that is, relevant) and one

[23] The first criterion is necessary to assess the second and not other way around.

Table 6.2 The impact of energy policy on access to electricity

	B	*Std. Error*	*Sig*	*Exp (B)*	*95% Wald Confidence Interval for Exp (B)*	
					Lower	*Upper*
Access to electricity = 0	0.8	0.45	0.09	2.15	0.88	5.24
Access to electricity = 1	3	0.63	0	20.1	5.89	68.51
Extensive body of energy legislation and regulation	1.3	0.35	0	3.81	1.93	7.53
Dispersed body of energy legislation and regulation	0	–	–	1	–	–
Community energy regulation	2.2	0.59	0	8.76	2.79	27.56
Model Fitting (sig.)	0					
Goodness-of-Fit, Pearson (sig.)	0.315					
Goodness-of-Fit, Deviance (sig.)	0.196					
Nagelkerke R square	0.429					
Test of Parallel Lines (sig.)	0.196					
Number of observations (n)	66					

is not. They confirm that having an extensive body of energy legislation and regulation expands access to electricity more than having a less extensive body (*H*1). For example, Mozambique has both an extensive body of energy legislation and regulation and has seen access to electricity progressing mainly from 2010. The results also confirm that being a pioneer in regulating community energy contributes to increased access to electricity more than being a lagger in regulating community energy (*H3*). This reflects the case of Ethiopia, which is the pioneer in regulating community energy and ranks first in access to electricity.

Having an extensive body of energy legislation and regulation (Mozambique) contributes to increasing access to electricity, but the effect of pioneering community energy regulation (Ethiopia) is of a greater magnitude. Having a dispersed or concentrated body of energy legislation and regulation has no impact at all on access to electricity (*H2*) as it is not significant. The model could not produce a meaningful B coefficient and significance. While the small number of observations may have affected the analysis, the results suggest that the key variables are extension and being a groundbreaker in energy policy rather than concentrating efforts at a particular point in time.

6.7 Discussion

Reflecting on Ethiopia being the pioneer in regulating community energy, its share of access to electricity through off-grid systems still remains low (11%) compared to on-grid (33%).[24] More than three-quarters of Ethiopians are rural dwellers,[25] and rural access to electricity is limited (26%).[26] Establishing small-scale off-grid systems in rural and hard-to-reach communities is still a challenge, regardless of Ethiopia being the first to regulate community energy through the legal arrangements mentioned above. The Government of Ethiopia is committed to increasing off-grid electricity access to 35% by 2025,[27] but most of Ethiopia's ***public investment in energy*** is concentrated on on-grid hydroelectric plants. While Ethiopia has introduced attractive package incentives and tax reliefs for private investment, issued licenses for electricity supply to the grid, regulated energy investments,[28] and set out the areas that foreign investors can invest in,[29] one might expect that the government would place more effort in attracting private and foreign investment in the energy sector. An influx of investment in energy projects enhances access to energy (Chirambo, 2016; ESMAP, 2022; McCollum et al., 2018; Menyeh, 2021; Pueyo, 2018; Sovacool, 2013). Access to energy requires improved not only financial mechanisms, such as a pro-poor public–private partnership model (Sovacool, 2013), but also substantial volumes of financing to provide energy for all by 2030 (Chirambo, 2016; Menyeh, 2021). In Africa, while private and public investments are insufficient to finance energy projects, public funding remains the primary source of investment in the energy sector (Baumli & Jamasb, 2020). This is the case of Ethiopia. Ethiopia leads in ***public investment in energy,*** followed by

[24] https://www.trade.gov/country-commercial-guides/ethiopia-energy#:~:text=According%20to%20the%20GOE's%20recently,achieve%20100%25%20electrification%20in%202025, retrieved June 29, 2023.

[25] https://tradingeconomics.com/ethiopia/rural-population-percent-of-total-population-wb-data.html, retrieved June 29, 2023.

[26] https://www.se4all-africa.org/seforall-in-africa/country-data/ethiopia/#:~:text=Ethiopia%20has%20a%20rich%20endowment,very%20limited%20throughout%20the%20country, retrieved June 29, 2023.

[27] https://tradingeconomics.com/ethiopia/rural-population-percent-of-total-population-wb-data.html, retrieved June 29, 2023.

[28] Proclamation 1180/2020.

[29] Regulation 474/2020.

Mozambique. Mozambique leads in *energy investment with private participation*.[30] However, the impulse of early adoption of off-grid regulations may have created confidence for investors and facilitated a wide range of interventions that translated indirectly into increased electricity access, even though this is not reflected in the rapid growth of community energy or other off-grid alternatives.

Malawi commenced regulating community energy two years earlier than Mozambique in 2019. This included the solicitation process and requirement for approval of mini-grid projects, grid-connection, tariff guidelines, system design, standard of compliance, licensing requirement and application of less onerous regulation (MERA, 2020). Yet Malawi lags on access to electricity, as it is lower on both public and private investment in energy. It might be expected that the Government of Malawi would allocate more public finance to energy access and would make efforts to attract more private and foreign investment in energy. The existing off-grid systems are challenged by a lack of affordability of electricity. About 62% of Malawi's population is multidimensionally poor (NSO, 2021), suggesting that where off-grid systems exist, poor rural Malawians tend not to make a connection.

Mozambique's 2021 community energy regulation has not been put to the test yet. Potential barriers include "the transparency of roles and responsibilities of key government institutions and having appropriate resource mechanisms in place—e.g., financial incentives or human capital, to facilitate the regulation's implementation" (Howe & Shenga, 2023). There are also 27 accessory instruments associated with the regulation to be published by the Government of Mozambique (Howe & Shenga, 2023), of which only five have been published in the Government Gazette. Delays in putting the regulation into practice and/or providing fiscal incentives to community energy operators may prevent rural access to electricity as investors may shift to invest in other countries or operators may find the energy business not profitable (see Baumli & Jamasb, 2020; ESMAP, 2022). Yet, Mozambique ranks first on investment in energy with private participation and second on public investment in energy. The extensive body of energy legislation and regulation, in general, is equally

[30] The World Bank Governance Indicators, https://databank.worldbank.org/source/world-development-indicators, retrieved November 14, 2022.

important in the Mozambique case, which seems to be paying off as electricity access increases in the country despite the late adoption of clearly beneficial alternatives.

Besides discussing the effect of investment on access to electricity, it is also important to reflect on technology, as investment is positively associated with technology, which also enhances access to electricity (Sovacool, 2012, 2013). Investment creates and transfers general knowledge and technologies in production and distribution (Osano & Koine, 2016). This suggests that countries that have high public (Ethiopia) and private (Mozambique) investments in energy may attract energy technologies. The World Economic Forum data indicates that the *technological readiness* and *innovation* of the three countries analysed are low, falling below the mid-point of 3.5.[31] Energy investment may be wasted and/or diverted to private gains in poor governance systems, undermining the impact of energy policies.

Strong governance is vital for developing policies to promote renewable energy consumption (Mahmood et al., 2021). In Africa, where governance systems are weak, corrupt, unstable and lacking in accountability and transparency, governance has been ineffective in improving access to electricity (Acheampong et al., 2022), irrespective of the adoption of energy policies. In the three countries, *government effectiveness* is very weak.[32]

In sum, universal access to electricity requires an extensive body of energy policy in general and regulation of community energy. Nonetheless, attracting private and public investments in energy, developing energy technologies and having an effective government system are equally important. Regulation alone will not be effective in articulating a just transition.

[31] World Economic Forum data.

[32] The Worldwide Governance Indicators, 2022 update. Aggregate governance indicators 1996–2021, *government effectiveness* "reflects perceptions of the quality of public services, the quality of the civil service and the degree of its independence from political pressures, the quality of policy formulation and implementation, and the credibility of the government's commitment to such policies". www.govindicators.org, retrieved January 16, 2024.

6.8 Conclusion

This chapter probed the effects of energy policies on access to electricity, comparing Ethiopia, Malawi and Mozambique by using longitudinal quantitative data tested through ordinal multiple regression analysis. The analysis suggests that, in general, more regulation results in increased rates of energy access, but this conclusion must be moderated by specifying that more regulation alone is not sufficient. Moreover, pioneering community energy correlates with increasing energy access, but this access is not necessarily due to off-grid energy. Thus, there is a need to investigate the causal links that explain these correlations. For example, a pioneering off-grid policy may be an indication of a proactive energy policy communities advancing off-grid proposals among other efforts, rather than pointing to an increase of off-grid of infrastructures alone.

The analysis confirms that regulating community energy and having an extensive body of energy legislation and regulation enhances access to electricity. The adoption of community energy and off-grid legislation may be simply a marker of a dynamic institutional context. However, there are challenges in promoting access to electricity through community energy regulation. Putting all the countries together, political governance is critical. Irrespective of being the pioneer regulating community energy (Ethiopia), having an extensive body of energy legislation and regulation (Mozambique), having community energy regulation approved fully (Malawi) and high public (Ethiopia) and private (Mozambique) investment in energy, all these have limited effect on access to electricity if governance systems are not improved. Money can be simply wasted and/or diverted to other ends, resulting in poor creation and transference of energy technologies.

In the case of Ethiopia, regardless of being a pioneer in regulating community energy, its huge public investment in energy has been primarily directed to on-grid energy systems rather than off-grid to a population that is more than three-quarters rural with very limited access to electricity. The rural communities where the grid system hardly reaches will continue to lack access to electricity in the coming years regardless of the Government's commitment.

In Malawi, although the Government has regulated community energy and introduced respective accessory instruments in 2019, the country lags on the public and private investment in energy that is necessary to create and transfer energy technologies to boost access to electricity. The few existing rural off-grid initiatives struggle to survive due to poverty, which inhibits most rural Malawians from affording electricity.

Mozambique leads in investment in energy with private participation, has an extensive body of energy legislation and regulation and comes second in public investment in energy. However, the country has not yet completed regulating community energy. Its community energy regulation was only recently adopted in December 2021, and as of May 2023, the Government has only published 5 out of 27 accessory instruments of the regulation. This suggests that there will be delays in creating and transferring general knowledge and technologies on off-grid systems consequently reducing energy poverty.

References

Acheampong, A. O., Shahbaz, M., Dzator, J., & Jiao, Z. (2022). Effects of income inequality and governance on energy poverty alleviation: Implications for sustainable development policy. *Utilities Policy, 78*(October), 101403. https://doi.org/10.1016/j.jup.2022.101403

Baumli, K., & Jamasb, T. (2020). Assessing Private investment in African renewable energy infrastructure: A multi-criteria decision analysis approach. *Sustainability, 12*(22), 9425. https://doi.org/10.3390/su12229425

Chirambo, D. (2016). Addressing the renewable energy financing gap in Africa to promote universal energy access: Integrated renewable energy financing in Malawi. *Renewable and Sustainable Energy Reviews, 62*(September), 793–803. https://doi.org/10.1016/j.rser.2016.05.046

Dava, F., & Tamele, V. (2011). *História dos 30 Anos da Electricidade de Moçambique, E.P.* Electricidade de Moçambique, E.P.

EDM. (2017). *Relatório Anual de Estatística 2015*. Electricidade de Moçambique, E.P.

ESMAP, Energy Sector Management Assistance Program. (2022). *Regulatory indicators for sustainable energy (RISE)*. World Bank.

Frelimo. (1977). *Directivas Económicas e Sociais. Documentos do 3° Congresso da Frelimo*. Partido Frelimo.

Howe, L., & Shenga, C. (2023). *Mozambique's off-grid energy regulation: Opportunities and challenges for the uptake of community energy projects*. CESET (Community Energy and the Sustainable Energy Transition).

Huhta, K. (2022). The contribution of energy law to the energy transition and energy research. *Global Environmental Change, 73*(March), 102454. https://doi.org/10.1016/j.gloenvcha.2021.102454

Kroth, V., Larcinese, V., & Wehner, J. (2016). A better life for all? Democratization and electrification in post-apartheid South Africa. *The Journal of Politics, 78*(3), 774–788. https://doi.org/10.1086/685451

Landman, T. (2009). *Issues and Methods in Comparative Politics*. An Introduction, 3rd ediition, Routledge.

Landman, T., & Carvalho, E. (2017). *Issues and methods in comparative politics*. Routledge.

Mahmood, H., Tanveer, M., & Furqan, M. (2021). Rule of law, corruption control, governance, and economic growth in managing renewable and nonrenewable energy consumption in South Asia. *International Journal of Environmental Research and Public Health, 18*(20), 10637. https://doi.org/10.3390/ijerph182010637

McCollum, D. L., Zhou, W., Bertram, C., De Boer, H. S., Bosetti, V., Busch, S., Després, J., Drouet, L., Emmerling, J., Fay, M., & Fricko, O. (2018). Energy investment needs for fulfilling the Paris Agreement and achieving the Sustainable Development Goals. *Nature Energy, 3*, 589–599. https://doi.org/10.1038/s41560-018-0179-z

Menyeh, B. O. (2021). Financing electricity access in Africa: A choice experiment study of household investor preferences for renewable energy investments in Ghana. *Renewable and Sustainable Energy Reviews, 146*(August), 111132. https://doi.org/10.1016/j.rser.2021.111132

MERA. (2020, July). *Regulatory framework for Mini-grids*. Malawi Energy Regulatory Authority.

Min, B. (2015). *Power and the vote: Elections and electricity in the developing world*. Cambridge University Press.

NSO. (2021). *Malawi Multidimensional Poverty Index Report 2021*. United Nations Development Programme and Oxford Poverty and Human Development Initiative.

Osano, H. M., & Koine, P. W. (2016). Role of foreign direct investment on technology transfer and economic growth in Kenya: A case of the energy sector. *Journal of Innovation and Entrepreneurship, 5*(31). https://doi.org/10.1186/s13731-016-0059-3

Przerworski, A., & Tenue, H. (1970). *The Logic of Comparative Social Inquiry*. Wiley.

Pueyo, A. (2018). What constrains renewable energy investment in Sub-Saharan Africa? A comparison of Kenya and Ghana. *World Development, 109*, 85–100.

Sovacool, B. (2012). The political economy of energy poverty: A review of key challenges. *Energy for Sustainable Development, 16*(3), 272–282.

Sovacool, B. (2013, December 1). Energy access and energy security in Asia and the Pacific. *Asian Development Bank Economics Working Paper Series*, 383. https://doi.org/10.2139/ssrn.2479159

To, L. S. (2015). *Renewable energy projects in rural China: A systemic capacity development approach* (PhD thesis). University of New South Wales.
World Bank. (2015). *Mozambique energy sector policy note*. World Bank.

CHAPTER 7

The Role of Community Energy and the Challenges in a State-Led Model of Service Provision in Ethiopia

Mulualem G. Gebreslassie

Abstract Community energy can drive sustainable energy transitions in Africa and beyond. However, the implementation of community energy systems is lagging because of the lack of appropriate governance frameworks. This paper aims to explore the critical challenges related to the governance and development of community energy systems in the context of state-led energy service provisions in Ethiopia and to recommend interventions to facilitate their implementation. The paper presents a systematic review of official energy policies, proclamations, and regulations documents, national and international publications, and a consultation with local energy agencies. The findings indicate huge gaps in energy governance, including technical, financial, and operational challenges.

M. G. Gebreslassie (✉)
Center of Energy, Ethiopian Institute of Technology-Mekelle, Mekelle University, Mekelle, Ethiopia
e-mail: mulualem.gebregiorgis@mu.edu.et

Materials and Engineering Research Institute, Sheffield Hallam University, Sheffield, UK

V. Castán Broto (ed.), *Community Energy and Sustainable Energy Transitions*, https://doi.org/10.1007/978-3-031-57938-7_7

Policymakers, the research community, and operators can take action to overcome these challenges. Strong commitment from all levels of government, international development organisations, and the private sector can make a difference in Ethiopia's community energy development. Dividing responsibilities for planning and implementing community energy is one critical step towards practical community interventions that can accelerate energy access, provide more reliable and affordable energy services, and meet the growing energy demand.

Keywords Community energy · Governance · Development, · Challenges · Electrification plans

7.1 Introduction

The world is experiencing a transition to the development and exploitation of renewable energy resources. This transition is driven by economies' dependence on the fluctuating cost of fossil fuels and the impact of fossil fuel exploitation on the environment (Bouzarovski & Tirado Herrero 2017; Evans & Phelan, 2016). Transitions are also driven by communities' social and behavioural transformations (Murphy, 2008). Community energy systems change energy governance from local energy resources because they provide models for citizens to own and manage energy systems. Community energy development commonly involves the exploitation of decentralised local renewable energy resources, ensuring lower carbon emissions and reduced transmission losses (Kiamba et al., 2022).

Community energy is growing in many parts of the world and is not just advanced in the context of international development. For example, the European Union member states acknowledge the contribution of community energy developments for sustainable energy transitions and are advocating for its development with greater participation of citizens (Fernandez, 2021). Examples include community energy projects in Denmark, Germany, the United Kingdom, and the Netherlands driven by renewable energy resources (Klein & Coffey, 2016; Thellufsen & Lund, 2016), contributing to sustainable energy transitions in these countries. These developments are driven by relatively better policies, incentive packages, and awareness of the advantages of collective action (Romero-Rubio & Díaz, 2015; Walker et al., 2010). Mini-grids are also cheaper and quicker tools to electrify remote areas where lower densities make extending networks unfeasible (Zebra et al., 2021).

In Ethiopia, the transition to renewable energy is comparatively faster in some rural areas (Erdiwansyah et al., 2021). Countries with low electricity access rates also have unaffordable, non-reliable energy infrastructures. Renewable energy-based off-grid systems are believed to be cost-effective in supplying power to remote areas (Bhuiyan et al., 2011). Off-grid energy developments could play a key role in bridging this energy access gap, but renewable energy still needs to be developed in the countries that need it the most. One of the reasons that prevents the development of off-grid and renewable energy is the need for suitable governance systems.

Ethiopia has seen massive changes in the energy landscape in the last three decades (Gebreslassie et al., 2022). One of the fundamental changes is the transition from a state-led model of a dominant centralised system to the development of integrated centralised and off-grid energy systems. This is driven by a combination of technological and institutional changes (Vargas & Davis, 2016) and the interest of communities and citizens to join in developing local energy resources. These changes offer an opportunity to develop renewable and off-grid energy systems that can support the Ethiopian government's ambitions of universal electricity access.

Different governments in Ethiopia have been putting effort into developing and changing energy governance. Still, these efforts have yet to deliver a full-scale development of off-grid systems (further compromised by the ravishing impacts of civil war). This paper asks what governance and development challenges prevent the development of community energy systems in Ethiopia.

The paper analyses current energy governance arrangements in Ethiopia by analysing policies, proclamations, regulations, and the legal basis of existing community energy initiatives. In doing so, it provides a snapshot of a moment in time, 2023, in which the transition may radically transform the landscape of energy access. An analysis of challenges in other developed and developing economies provides a reference for comparative analysis for developing measures that could help the development of community energy.

7.2 Community Energy in the Ethiopian Context

Klein and Coffey (2016) consider community energy as any energy project that could serve anything from a single household to medium-sized systems and that could be managed locally through institutions such as cooperatives. Röder et al. (2017) and Rogers et al. (2008) define community energy as a decentralised system developed by high community involvement in the development process for their consumption. The IRENA Coalition for Action defined community energy as "*the economic and operational participation and ownership by citizens or members of a defined community in a renewable energy project*" (IRENA Coalition for Action, 2018). Community energy is also defined as collectively owned local systems (Wirth, 2014).

In the Ethiopian context, community energy is not defined in energy policy, regulations, or investment proclamations. However, official documents in Ethiopia distinguish between national grid and off-grid energy systems. In practical terms, community energy is categorised as part of the off-grid energy systems, including standalone and mini-grid systems. Standalone energy systems serve single customers such as households and health and education institutions, while mini-grid systems serve a group of households that form a community. The current mini-grid systems, serving tens and hundreds of households, include solar-based mini-grid systems, micro-hydropower systems, and hybrid systems, grouped under the generic name of off-grid energy systems, even when communities own them.

In line with the literature definitions and current regulations, in Ethiopia, community energy systems are defined as mini-grid systems owned, managed, operated, or used by a group of households that constitute a community.

7.3 Off-Grid Energy Governance in Ethiopia

7.3.1 Existing Policy

The current Ethiopian Energy Policy was issued in 1994, during the period of the Transitional Government of Ethiopia. This was the first attempt for a policy document to consider the wide-ranging concerns of all sectors. The 1994 energy policy was conceived in the shadows of the country's flagship economic development strategy, the Agricultural Development Led Industrialization (ADLI). The ADLI placed agriculture as the economy's driving force and envisaged the Ethiopian economy's structural transformation through export-led growth, which feeds into an interdependent agricultural and industrial development.

The policy objectives focused on critical areas for transforming the county's economic and social goals. These consisted of giving high priority to state-led centralised hydropower resource development, as hydrological resources are Ethiopia's most abundant and sustainable energy forms; implementing appropriate policy measures to achieve a gradual transition from traditional energy fuels to modern fuels; ensuring energy efficiency through setting issuing and publicising standards and codes; developing human resources and competent energy institutions; providing the private sector with necessary support and incentives to participate in the development of the energy resources, and incorporating environmental protection in energy production and use.

A key focus was the involvement of the private sector in the development of energy resources, which could have facilitated community energy developments. However, community energy systems have yet to materialise, though few initiatives have started recently. The government, private sectors, cooperatives and the public could be involved in community energy.

When the policy was developed, it was intended to be dynamic—allowing for revisions occasionally to incorporate new developments in the sector and the broader economy. However, the policy was never updated.

An attempt to draft a new document, the "Ethiopian National Energy Policy," occurred in 2013 (MoWIE, 2013). This draft indicated the need to update the existing policy to respond to structural and transformational changes at national and international levels. Some of the key policy directions that were not included in the existing policy but envisioned to be included in the new policy document consisted of:

- Developing and using all renewable energy resources, including biofuels;
- Addressing the development and use of new technologies such as electric rail, electric cars, hybrid cars, and flex-fuel vehicles;
- Providing great attention to energy efficiency at the supply and demand side where there are substantial power losses;
- Promoting the localisation of technologies to minimise dependence on imported technologies and to reduce energy development project costs;

- Aligning the policy with the Climate Resilient Green Economy (CRGE) strategy being implemented in the country;
- Promoting regional integration to get foreign currency and to play a role in regional geopolitical stability;
- Foreseeing the development of nuclear energy for power generation in the future;
- Implementing measures for a transition from fuel wood to modern energy to reduce deforestation;
- Considering cross-cutting issues such as energy regulatory frameworks, energy sector governance, building vital energy institutions and capacity, integrated energy planning, energy efficiency and conservation, energy pricing, research and development, environmental and social impact, gender, and regional and international cooperation.

These additions could be critical for transforming Ethiopia's energy landscape and promoting the massive development of community energy systems. The draft policy contains substantial and essential policy instruments to support the effective implementation of the policy objectives. Addressing cross-cutting issues may change the energy sector's ineffective governance and development to support the country's ambitious economic development plans. However, the civil war has delayed the policy development and implementation in every sector, including energy.

7.3.2 Energy Proclamations and Regulations

The Energy Regulation No. 447/2019 (Council of Ministers, 2019) outlines how to obtain generation, transmission, distribution, and sales licenses as well as licenses for importing and exporting electricity. The regulation provides clear rights and obligations for these licensees and customers. The duration of the licenses differs according to the type of power resources, with hydro and geothermal power generation up to 25 years; wind, solar, biomass, urban waste, thermal, and biogas power generation up to 20 years; transmission up to 30 years; and distribution and sales up to 20 years.

The Ethiopian Energy Authority (EEA) is the sole regulator for government and private energy developments. The authority has licensing and regulatory mandates for the generation, transmission, distribution, and sales within the power sector. The EEA defines and enforces licensing requirements, parties' rights and obligations, amendments, and certificates of competency. It also advises the Government on tariff proposals for both grid and off-grid energy systems (Council of Ministers, 2019).

Though the private sector is allowed to generate electricity, the Investment Proclamation No. 280/2002 (Council of Ministers, 2002) reserves the transmission and distribution of electricity as areas to be targeted through public investment or joint ventures with the government. Decisions on the electricity tariffs are still in the hands of the state-owned authority, which also holds exclusive rights for the transmission, distribution, and supply of electricity through the integrated national grid system. Consequently, the participation of the private sector has only been limited to Engineering, Procurement, and Construction (EPC) contracts. However, the private actors are given full responsibility to engage in off-grid generation, transmission, distribution, and sales of electricity, though the tariff is still decided by the state-owned authority in consultation with both the private sector and customers.

The regulation provides guidance for grid integration when the grid expansion overlaps with off-grid energy systems. Still, it lacks clarity on how to proceed if the off-grid system is unsuitable for grid integration.

The Energy Proclamation obliges transmission and distribution network owners (Ethiopian Electric Power (EEP) and Ethiopian Electric Utility (EEU)) to give access to other license holders after payment of a prescribed fee (Council of Ministers, 2013). This protection is important for Independent Power Producers (IPPs) because the electricity market is not fully unbundled, and the sole transmission network operator (EEP) is also active in power generation. The enactment of a third-party access rule can provide a guarantee for IPPs, ensuring that the transmission network operator will connect to all generation facilities without unduly favouring generation facilities owned and operated by the state-owned enterprise (EEP).

Several regulations support an enabling ecosystem to facilitate off-grid connections, including harmonised national quality standards (i.e. in agreement with the Global Lighting standards), an exemption of both duty and excise for importers of solar products under 15Wp, and lifting of testing (Pre-Verification of Conformity is sufficient). Microfinance institutions (MFIs) and private sector enterprises (PSEs) have been critical stakeholders in facilitating and supplying solar lighting, charging products, and solar home systems through a credit line at the Development Bank of Ethiopia (DBE). DBE is mandated to facilitate the flow of funding the World Bank provides for energy development. The following specific products are eligible under the credit line at the DBE: Solar lanterns, solar home systems, solar water pumps, mini-grid developments (solar, wind, biomass, diesel), waste-to-energy technologies, improved cook stoves, domestic biogas, and ethanol stoves. With this credit line, capacity-building activities are provided to the MFIs and PSEs in collaboration with the regional energy bureau to ensure proper implementation.

7.4 Off-Grid Energy Developments in Ethiopia

7.4.1 Evolving Electrification Plans

7.4.1.1 Universal Electrification Access Programme (UEAP)

The Universal Electrification Access Programme (UEAP) was launched by the government of Ethiopia in 2005 (MoWIE, 2019) with the objective of expanding the state-led model grid network to 60% of towns and villages and increasing the number of EEU customers from 800,000 to about 2.8 million from 2005 to 2015. Following this experience, the National Electrification Strategy (NES) was developed in 2016 with critical recommendations for new electrification plans. The NES was the starting point for the integrated National Electrification Programme 2.0 (NEP 2.0), whose implementation started in 2019. As part of the development of the NEP, a Multi-Tier Framework (MTF) household survey was conducted in 2017–2018 to map the electrification coverage. The access rate for the country overall was 44%, 33% provided through the grid, and 11% through off-grid connections, excluding the provision of lantern technologies. About 56% of the population lacked access to an adequate and reliable source of electricity (MoWIE, 2019). The survey also points out lower connectivity in rural areas as the main challenge for the poor access outlook of the country. The success and failures of the previous electrification plans and recent national and international transformations and dynamics of the energy sector have provided a foundation for the next electrification plan by diversifying its scope into the off-grid energy systems.

7.4.1.2 *National Electrification Programme 2.0: Integrated Planning for Universal Access*

The second national electrification programme (NEP 2.0) continues the previous plan, adding detailed planning, policies, and regulations to overcome the challenges encountered in the last electrification plan. NEP aims to achieve 100% electrification by 2025 while extending the grid to 96% grid connections by 2030. NEP 2.0 intends to facilitate government commitment to a technically sound geospatial least-cost rollout plan sustained by comprehensive institutional arrangements and a sustainable financing platform.

The Ministry of Water, Irrigation and Energy (MoWIE) has developed the NEP implementation framework identifying the responsibilities for both the grid and off-grid system developments. MoWIE leads the development of NEP's enabling policies with the support of development partners. The Ethiopian Energy Authority (EEA) is the sole accountable body for approving required technical standards and regulations, which means that the regional Ethiopian Energy Utility (EEU) branches are not able to draft their standards. This has multiple political and economic ramifications, including challenging the regional EEU branches' reliance on local manufacturing. At the same time, the power to issue licenses is partially delegated to regional EEU branches, but they are not always able to action it. EEU branches, cooperatives, and businesses within the private sector may develop off-grid energy systems. There is an assumption of full coordination among implementing actors at all levels, leveraging on their respective experiences and capacities, but this ideal model overlooks the difficulties inherent to the establishment and sustainability of institutions, their diverging interests, or the limited communication between them. Thus, there is a tension between the institutional order imagined in NEP 2.0 and the realities of action on the ground. Institutional rearrangement is, nevertheless, the essential means of change in NEP 2.0.

Expanding both on-grid and off-grid systems is core to the NEP 2.0 objectives, with 65% of the population connected through the state-led centralised system services and the rest through off-grid systems. Improving connectivity requires territorial strategies of densification (≤1 km), intensification (1–2.5 km), and extension (2.5–25 km) (MoWIE, 2019).

In addition, a considerable investment is required. NEP 2.0 total costs are estimated at US$6 billion, 42% of which should be directed towards off-grid systems. The national electrification programme intends to leverage public, private, and developmental partners (DPs) finances. For example, the Government is expected to contribute an estimated 40% (US$1 billion) of the cost of off-grid development, and the remaining 60% (US$1.5 billion) to be covered by DPs and private sector resources (MoWIE, 2019).

In November 2020, a civil war erupted in Ethiopia, which directly impacted the implementation of NEP 2.0. The war devastated the country's economic activity, leading to the suspension of programmes and projects due to security, shortage of resources, and lack of sustained commitment from the government. The impact of the war is continuing, and the future of NEP 2.0 is still compromised.

7.4.2 *Progress of Off-Grid Energy Deployments*

A government survey investigating the expansion of the off-grid energy systems recorded 2.2 million connections by the end of 2018 (MoWIE, 2019), involving public and private sectors and a heterogeneous set of funding mechanisms.

7.4.2.1 Solar Home Systems and Lanterns

Massive expansions of solar lighting were initiated in 2003 by establishing public funding mechanisms, such as the Rural Electrification Fund (REF). REF has supported the deployment of solar PVs for nearly 45,000 solar home systems, 1,049 schools, 1,576 health posts, and 696 rural electric service cooperatives. In addition, the private sector has been heavily involved in the deployment of solar lighting and solar lanterns. A World Bank and the DBE partnership enabled a two-phase credit facility, the Market Development Credit Line (MDCL), from 2012 to 2019 to finance standalone solar projects. The MDCL supported private actors to deploy nearly 1,124,838 solar lanterns and 72,370 solar home systems from 2012 until 2018 (MoWIE, 2019).

7.4.2.2 Community Energy Initiatives

Community energy development has been ongoing in Ethiopia for some time, supported by technical and financial support from NGOs, the private sector, government, and development partners. The Ethiopian Electric Utility owns 31 isolated diesel-based mini-grid systems with a capacity ranging from 100 kW to 520 kW. These mini-grid systems connect nearly 8,000 customers (MoWIE, 2019). The systems are operated by the EEU, and in the event of grid arrival, these systems are easily connected to the national grid. As part of their plan, they have already connected five diesel-based mini-grid systems to the grid and will do it for the remaining 31 mini-grids in the event of overlapping.

Under the new NEP 2.0, 13 mini-grid systems have been initiated since 2019, with a capacity ranging from 6 kW to 52 kW from hydro, solar, and wind resources. The development partners supporting these initiatives include GIZ/EnDev, Power Africa, and the United Kingdom's Department for International Development (DFID), with a combination of private and cooperative business and institutional models. The progress of these initiatives is not clear, considering the conflict that has ravishing the country since November 2020.

A few mini-grid energy systems owned by the community have also been developed based on the two electrification plans. Accordingly, 17 community energy systems have been developed in Ethiopia. Most of these energy systems are based on hydro, solar, and one hybrid with diesel energy resources, as indicated in Table 7.1. These initiatives are mainly based in three regions of Ethiopia, namely Tigray, Southern Nations Nationalities People's Region (SNNPR), and Oromia.

According to the regional Mining and Energy Agency, the solar-based mini-grid system found in Tigray, inaugurated in 2020, has a capacity of 20.4 kW serving nearly 200 households. This system has state-of-the-art technologies to control each household's energy consumption remotely. The communities use the system for lighting, television, and charging, but they still use traditional fuels for energy-intensive cooking and baking applications.

The Oromia region has seven community energy systems based on the information obtained from the regional energy agency. Six of these energy systems are hydro-based and one is a hybrid of solar and diesel-based

Table 7.1 Community energy initiatives in Ethiopia

Region	*Type of energy*	*Number of community energy systems*
Tigray	Solar mini-grid	1
SNNPR	All Hydro	9
Oromia	6-Hydro, one solar-diesel hybrid	7

Table 7.2 Micro hydropower community energy schemes in SNNPR

No	*Name of Zone*	*Name of a specific area*	*Status*
1	Sidama zone	Keramo	Functional
2	Sidama zone	Murago	Non-functional
3	Sidama zone	Allo	Functional
4	Sidama zone	Gobecho 1	Non-functional
5	Sidama zone	Gobecho 2	Non-functional
6	Sidama zone	Hagere Sodicha	Functional
7	Sidama zone	Erertie	Functional
8	Gedeo zone	Rago Senbete	Non-functional
9	Gedeo zone	Rasa Dango	Non-functional

energy resources. Similarly, nine community energy systems are derived from hydro in SNNPR, five of which are non-functional (55.6%), and the rest are functional, as shown in Table 7.2.

7.5 Community Energy Governance and Development Challenges

7.5.1 Challenges of the Energy Governance

Institutional flaws such as weak policies, regulatory frameworks, and lack of strong institutions hamper the deployment of decentralised community energy systems (Feron, 2016). The developed world, such as the European Union, is still lagging in the development of a conducive regulatory framework to make the community energy systems competitive with large utility companies, though there are encouraging developments in some of the member states (Fernandez, 2021). The regulation favours large companies affecting the small community energy initiatives to compete in the tendering process. This is a clear indication that developing a suitable regulatory framework is a challenge for most countries, as it requires a learning time on how to make the development competent and attractive for any stakeholders involved in the sector.

In Ethiopia, the lack of timely policy revisions has a huge impact on the development of community energy systems. There is a lack of clear policy instruments that support the effective implementation of the outlined policy objectives, a lack of clear legal and administration directions from the policy document on how to manage the community energy systems, and weak community involvement in the development and decision-making process such as planning, engineering work, operation, and management. Similar studies in two existing mini-grid systems in Namibia indicated that the lack of community involvement in the design phase combined with the lack of standards and suitable regulations pose huge challenges for the implementation of off-grid energy systems (Hoeck et al., 2022). A study by Lennon et al. (2019) proved that the lack of greater participation of the community in the development and decision-making process is a source of resistance and frustration affecting the effectiveness of community energy developments and their sustainability.

There is an overlapping of the grid and off-grid energy systems and thus the lack of clear grid arrival policy is a critical barrier for investments in community energy systems (IRENA, 2019). The current energy regulation in Ethiopia lacks clarity on how to proceed if community energy systems are not suitable for grid integration. This has a huge impact on the involvement of the private sector to develop community energy. Coupled with the lack of a precise definition of community energy in the current energy policy, the lack of procedural clarity poses barriers for the administration, operation, and maintenance of the systems.

Ethiopia lacks cost-reflective tariffs to attract the private sector and reduce the need for public support for community energy developments. Empirical studies show that grid-based electricity is cheaper compared to off-grid-based generation of electricity (Lukuyu et al., 2020 and Franz et al., 2014), which also tends to discourage the private sector from investing in community energy. Moreover, there are huge transaction costs, such as, for example, the need to obtain separate licenses for each of the energy supply chains (generation, transmission, distribution, and sales) regardless of size or location. In summary, the governance structure not only impacts the working of regulations but also the administrative arrangements required to make community energy possible.

7.5.2 Lack of Integrated Planning for Energy Development

Though NEP 2.0 stipulates the development and implementation of integrated planning, it is yet to be materialized. The development of community energy is being implemented without national integrated planning. When both off-grid and on-grid systems are available, prospective users tend to shift to the on-grid system: mini-grid systems are abandoned in the event of overlapping (Tenenbaum et al., 2018). At the moment, grid systems are perceived as being relatively better in terms of quality of services, price, and sustainability. Though energy regulations allow for the integration of community energy systems in the event of overlapping (provided that it is suitable for integration), there is no clear mechanism or enforcing laws to ensure that off-grid systems must conform to such settings. The integration of renewable energy systems into the grid is always challenging due to the changes created in the system inertia affecting the voltage and frequency control (Holjevac et al., 2021; Meliani et al., 2020), and thus careful design of the systems is required to maintain such option open.

7.5.3 Technical and Operational Challenges

Sustainability of service provision is also a challenge for community energy systems. The systems quickly malfunction after a service of a few years. Maintaining the system is often a challenge because many governance arrangements do not distinguish between developers, owners, and operators (Avelino et al., 2014). The lack of well-structured organizational, technical, and management systems, improper power demand assessment, limited knowledge about the development of projects, ownership ambiguities, and lack of maintenance capabilities further hinder the success of community energy projects (Madriz-Vargas et al., 2015).

Mini-grids are uniquely exposed to technical, operational, and financial risks (Keisang et al., 2021). Empirical evidence shows that, in practice, limited knowledge of the operation of the systems, high maintenance costs, and lack of spare parts are some of the main reasons why off-grid projects fail in Tigray, a regional state of Ethiopia (Gebreslassie, 2020). Lack of technical know-how and monitoring systems is compounded by a lack of technical standards to facilitate its use and maintenance, and limited possibilities to access education among the communities that are purported to receive the benefits of community energy (Hoeck et al., 2022). Financial returns are risky because community projects often serve areas with low electricity demand where people have limited ability to pay, and hence, profit margins are very low (Manetsgruber et al., 2015; USAID, 2018).

7.6 Enabling Governance that Supports Community Energy

The governance and development challenges described above constitute critical bottlenecks for community energy projects and require policy and regulatory interventions. Specific revisions could be made to the existing policy framework to remove some barriers that prevent community energy. First, there is a need to gain clarity in the regulation, for example, by establishing a clear definition of community energy that relates to the practice of community energy on the ground and that reflects the growing enthusiasm for cooperative models of energy management. Such definition needs to take into account the forms of social organisation in Ethiopia and the structures of social organization that they provide. In Ethiopia, this kind of sociological analysis still needs development but the possibility to develop a consensual definition of community energy among relevant stakeholders is within reach.

Second, licensing constitutes a major administrative barrier. As explained above, current licensing requirements increase the administrative burden for public authorities and it is time-consuming for community energy developers. This can be addressed through a capacity-based license requirement, something already piloted in other countries. For example, in Tanzania, generation licenses are not required for mini-grids with less than one MW capacity (IRENA, 2016). In Rwanda, the exemption is applied for projects with less than 50 kW capacity (Zebra et al., 2021). Those regulations also allow a single license for multiple sites. The current licensing regulation in Ethiopia requires separate licenses for generation, transmission, distribution, and sales while simultaneously does not specify if a developer can acquire full licensing for the whole energy supply chain. Single license for the whole supply chain could reduce the burden on the government and the licensee. Countries like Sierra Leone have already experienced using whole supply chain licenses (SEforALL, 2020).

Third, better integration assurances are required. The energy regulation in Ethiopia allows for the integration of community energy systems into the national grid in the event of overlapping if the system is suitable but it remains unspecified if integration is not possible. This requires addressing two challenges. First, there is a need to condition standards for off-grid development with a view to facilitate their future integration. Second, assurances must be provided for investors to ensure that projects will be retained or compensated in the event of overlapping (Antonanzas et al., 2021). For example, countries such as Nigeria have developed provisions to address the event of grid overlapping such as providing full compensation or enabling off-grid providers to continue getting revenue by allowing the operation of the off-grid system alongside the on-grid one (SEforALL, 2020).

Finally, addressing financial barriers is a key aspect of facilitating community energy. A Feed-in Tariff (FiT) is a means to encourage renewable development and can play a key role in supporting community energy projects that may face insurmountable barriers until the market is established, by providing a framework for negotiating power purchase agreements between producers, investors, and public authorities. FiTs encourage private sector involvement in community energy because they assure long-term purchase agreements for developers and reduce financial risks (Couture et al., 2010). For example, in Vietnam, the government has implemented FiTs since 2011 (Le et al., 2022). By the end of 2020, the country had reached 600 MW of installed wind power capacity and 17.6 GW of installed solar power capacity. A study of 30 OECD member countries in the period 1990–2011 demonstrated that FiT policies accelerated the deployment of solar PV (Dijkgraaf et al., 2018). Other countries like Algeria, Kenya, Mauritius, Rwanda, South Africa, Tanzania, Philippines, and Uganda have already implemented FiT policies subsequently seeing encouraging progress in the deployment of renewable energy-based off-grid energy systems (Lagac & Yap, 2021; Nganga et al., 2013). Based on different experiences of the existing FiT in different countries, payments are recommended to vary based on location, resource quality, project size, grid and off-grid systems, and type of technologies (Cox & Esterly, 2016).

Ethiopia is one of the African countries that has already developed FiT proposals but has not yet implemented them. NEP 2.0 proposes cost-reflective tariffs that are expected to be within a range from 0.30 US$/kWh to 1.20 US$/kWh based on international experiences, which is much higher than the current grid-based electricity tariff. This could attract the private sector to be involved in the development and support the government's ambition of universal electrification access, but the tariffs have yet to be approved and implemented. Energy and economic planning should go hand in hand to facilitate investors' confidence. For example, community energy can contribute to diversify the economy through productive applications, such as irrigation systems, which not only increases community confidence and sense of ownership but also facilitates investors' confidence (Chirambo, 2018).

Developing enabling regulations and providing resources are necessary but not sufficient preconditions for community energy developments (Ambole et al., 2021). Community energy also requires specific governance instruments addressing the need for participatory business models, innovative financing mechanisms, mechanisms for greater participation of communities, and incentive packages (e.g., tax incentives). Regulations must recognise community-based governance models, distinct from private-led or utility-led models prioritised in development plans (Pokhrel et al., 2013). Community energy depends on the establishment of participatory business models that provide communities with the opportunity to engage in the decision-making processes and invest in the development of community energy projects while simultaneously addressing the inequalities within the community that may hinder participatory energy governance (Lennon et al., 2019). This means that community energy requires distinct forms of support not currently contemplated in policies such as the NEP 2.0.

The success of community energy depends on achieving and maintaining a strong sense of ownership by the communities; to build community capability to manage, operate, and resolve conflicts; technology selection and sizing, to build strong capability and involvement of the community in the operation and maintenance; and to ensure energy service reliability (Madriz-Vargas et al., 2015). Greater community participation throughout the project development, capacitating the community, and developing workable governance with clear structures is essential for the sustainability of community energy projects (Gill-Wiehl et al., 2022; Katre et al., 2019). Solid and effective collaboration among stakeholders is also critical for mini-grid developments (Korkovelos et al., 2020). This requires coupling engineering and economic analysis with sociological analysis on the relationship between the energy transition and the social relations that make it possible.

7.7 Conclusion

In summary, there are important challenges that prevent community energy in Ethiopia. Some of these challenges may be administrative or technological, and can be addressed through relatively feasible regulatory reforms. However, many challenges relate to misunderstandings regarding the systems of provision of electricity and political factors that structure energy governance in the country. Policymaking in the energy sector has been characterised by a lack of timely policy revisions and unclear policy instruments, translating, for example, into ambiguous requirements for off-grid integration or increased investment risks. The institutional landscape is also not conducive to community energy development, sometimes directing my misunderstandings about the possibilities of community involvement or the prioritization of interests that favour large companies.

For the government's ambitious universal electrification programme to succeed, it is essential to actively engage in regulatory reform to implement internationally proven successful tools in developing and managing community energy systems. Facilitating licensing for small projects, developing a grid arrival policy, and implementing FiT proposals are feasible steps towards an enabling environment for community energy. Further, explicit recognition of community energy will facilitate planning that encourages its development.

The real challenge however lies in the means to devise mechanisms to facilitate greater participation of the community throughout the community energy development from inception to operation of the systems and facilitate the generations of business models that enable communities to manage community energy systems. Such intent raises complex demands including frameworks for conflict resolution, developing the community's operational capacity, and facilitating multi-stakeholder partnerships to support investment and facilitate project sustainability. Community energy calls for rethinking current energy systems in relation to their social function. Diversifying the knowledge base (Chapter 8) is essential. On the one hand, diversifying the knowledge base means that engineering knowledge should be complemented with social research skills to understand the heterogeneous systems of governance that will facilitate a transition to sustainable energy. On the other hand, diversifying the knowledge base also requires recognising the expertise held within communities and expanding it to facilitate their participation in the implementation and operationalisation of community energy. Community energy calls for nothing less than the democratisation of energy knowledge.

References

Ambole, A., Koranteng, K., Njoroge, P., & Luhangala, D. L. (2021). A review of energy communities in sub-Saharan Africa as a transition pathway to energy democracy. *Sustainability, 13*, 2128. https://doi.org/10.3390/su13042128

Antonanzas, F., Antonanzas, J., & Blanco-Fernandez, J. (2021). State-of-the art of mini grids for rural electrification in West Africa. *Energies, 14*, 990. https://doi.org/10.3390/en14040990

Avelino, F., Bosman, R., Frantzeskaki, N., Akerboom, S., Boontje, P., Hoffman, J., Paradies, G., Pel, B. Scholten, D., & Wittmayer, J. (2014). *The (self-) governance of community energy: Challenges & prospects*. Drift Practice Brief no. PB 2014.01, Rotterdam. Retrieved January 16, 2024, from https://drift.eur.nl/publications/the-self-governance-of-community-energy/

Bhuiyan, A. M., Islam, K. S., Haque, M. M., Mojumdar, M. R., & Rahman, M. M. (2011). Community-based convenient hybrid mini-grid: Implementation proposal & analysis for Bangladesh. *International Journal of Innovation, Management and Technology, 2*(5), 354.

Bouzarovski, S., & Tirado Herrero, S. (2017). The energy divide: Integrating energy transitions, regional inequalities and poverty trends in the European Union. *European Urban and Regional Studies, 24*(1), 69–86. https://doi.org/10.1177/0969776415596449

Chirambo, D. (2018). Towards the achievement of SDG 7 in sub-Saharan Africa: Creating synergies between Power Africa, sustainable energy for all and climate finance in order to achieve universal energy access before 2030. *Renewable and Sustainable Energy Reviews, 94*, 600–608.

Council of Ministers. (2002). *Investment Proclamation No. 280/2002*. Federal Negarit Gazette of the Federal Democratic Republic of Ethiopia.

Council of Ministers. (2013). *Energy Proclamation No. 810/2013*. Federal Negarit Gazette of the Federal Democratic Republic of Ethiopia.

Council of Ministers. (2019). *Energy Regulation No. 447/2019*. Federal Negarit Gazette of the Federal Democratic Republic of Ethiopia.

Couture, T. D., Cory, K., Kreycik, C., & Williams, E. (2010). *A policy makers' guide to feed in tariff policy*. National Renewable Energy Laboratory. Retrieved January 13, 2024, from http://www.nrel.gov/docs/fy10osti/44849.pdf

Cox, S., & Esterly, S. (2016). *Feed-in Tariffs: Good Practices and Design Considerations: A clean energy regulators initiative report*. National Renewable Energy Laboratory, Technical Report NREL/TP-6A20-65503. Retrieved January 13, 2024, from https://www.nrel.gov/docs/fy16osti/65503.pdf

Dijkgraaf, E., van Dorp, T. P., & Maasland, E. (2018). On the effectiveness of feed-in tariffs in the development of solar photovoltaics. *The Energy Journal, 39*(1). https://doi.org/10.5547/01956574.39.1.edij

Erdiwansyah, M., Zaki, M., Gani, A., & Muhibbuddin, J. (2021). A review of renewable energy mini-grid systems in the non-interconnected rural areas: A case study. *Journal of Hunan University (natural Sciences), 48*(1), 133–151.

Evans, G., & Phelan, L. (2016). Transition to a post-carbon society: Linking environmental justice and just transition discourses. *Energy Policy, 99*, 329–339.

Fernandez, R. (2021). Community renewable energy projects: The future of the sustainable energy transition? *The International Spectator, 56*(3), 87–104. https://doi.org/10.1080/03932729.2021.1959755

Feron, S. (2016). Sustainability of off-grid photovoltaic systems for rural electrification in developing countries: A review. *Sustainability, 8*(12), 1326. https://www.mdpi.com/2071-1050/8/12/1326

Franz, M., Hayek, N., Peterschmidt, N., Rohrer, M., Kondev, B., Adib, R., Cader, C., Carter, A., George, P., Gichungi, H., Hankins, M. (2014). *Mini-grid policy toolkit: Policy and business frameworks for successful mini-grid rollouts*. IAEA. Retrieved January 13, 2024, from https://inis.iaea.org/search/search.aspx?orig_q=RN:49053206

Gebreslassie, M. G. (2020). Solar home systems in Ethiopia: Sustainability challenges and policy directions. *Sustainable Energy Technologies and Assessments, 42*, 100880. https://doi.org/10.1016/j.seta.2020.100880

Gebreslassie, M., Cuvilas, C., Zalengera, C., To, L. S., Baptista, I., Robin, E., Bekele, G., Howe, L., Shenga, C., Macucule, D. A., Kirshner, J., Mulugetta, Y., Power, M., Robinson, S., Jones, D., & Castán Broto, V. (2022). Delivering an off-grid transition to sustainable energy in Ethiopia and Mozambique. *Energy, Sustainability and Society, 12*, 23. https://doi.org/10.1186/s13705-022-00348-2

Gill-Wiehl, A., Miles, S., Wu, J., & Kammen, D. M. (2022). Beyond customer acquisition: A comprehensive review of community participation in mini grid projects. *Renewable and Sustainable Energy Reviews, 153*, 111778.

Hoeck, I., Steurer, E., Dolunay, Ö., & Ileka, H. (2022). Challenges for off-grid electrification in rural areas. Assessment of the situation in Namibia using the examples of Gam and Tsumkwe. *Energy, Ecology and Environment, 7*, 508–522. https://doi.org/10.1007/s40974-021-00214-5

Holjevac, N., Baškarad, T., Đaković, J., Krpan, M., Zidar, M., & Kuzle, I. (2021). Challenges of high renewable energy sources integration in power systems: The case of Croatia. *Energies, 14*, 1047. https://doi.org/10.3390/en14041047

IRENA. (2016). *Policies and regulations for private sector renewable energy mini-grids*. Retrieved January 8, 2024, from http://www.irena.org/DocumentDownloads/Publications/IRENA_Policies_Regulations_minigrids_2016.pdf

IRENA Coalition for Action. (2018). *Community energy: Broadening the ownership of renewables*. International Renewable Energy Agency.

IRENA. (2019). *Off-grid renewable energy solutions to expand electricity access: an opportunity not to be missed*. International Renewable Energy Agency. Retrieved January 14, 2024, from https://www.irena.org/publications/2019/Jan/Off-grid-renewableenergy-solutions-to-expand-electricity-to-access-An-opportunity-not-to-be-missed

Katre, A., Tozzi, A., & Bhattacharyya, S. (2019). Sustainability of community-owned minigrids: Evidence from India. *Energy, Sustainability and Society, 9*(2), 1–17. https://doi.org/10.1186/s13705-018-0185-9

Keisang K., Bader, T., & Samikannu R. (2021). Review of operation and maintenance methodologies for solar photovoltaic micro-grids *Frontiers in Energy Research, 9*, 730230. https://www.frontiersin.org/articles/10.3389/fenrg.2021.730230/full

Kiamba, L., Rodrigues, L., Marsh, J., Naghiyev, E., Sumner, M., Empringham, L., De Lillo, L., & Gillott, M. (2022). Socio-economic benefits in community energy structures. *Sustainability, 14*(3), 1890. https://doi.org/10.3390/su14031890

Klein, S. J., & Coffey, S. (2016). Building a sustainable energy future, one community at a time. *Renewable and Sustainable Energy Reviews, 60*, 867–880.

Korkovelos, A., Zerri, H., Howells, M., Bazilian, M., Rogner, H. H., & Fuso Nerini, F. (2020). A retrospective analysis of energy access with a focus on the role of mini-grids. *Sustainability, 12*, 1793.

Lagac, J. M. P., & Yap, J. T. (2021). Evaluating the feed-in tariff policy in the Philippines. *International Journal of Energy Economics and Policy, 11*(4), 419–425.

Le, H. T. T., Sanseverino, E. R., Nguyen, D. Q., Di Silvestre, M. L., Favuzza, S., & Pham, M. H. (2022). Critical assessment of feed-in tariffs and solar photovoltaic development in Vietnam. *Energies, 15*, 556. https://doi.org/10.3390/en15020556

Lennon, B., Niall P. Dunphy, N. P., & Sanvicente, E. (2019). Community acceptability and the energy transition: a citizens' perspective. *Energy, Sustainability and Society, 9*(35). https://doi.org/10.1186/s13705-019-0218-z

Lukuyu J., Muhebwa A., & Taneja J. (2020, June 22–26). Fish and chips: Converting fishing boats for electric mobility to serve as minigrid anchor loads. *E-Energy '20: Proceedings of the Eleventh ACM International Conference on Future Energy Systems*. Virtual Event, Australia. https://doi.org/10.1145/3396851.3397687

Madriz-Vargas, R., Bruce, A., & Watt, M. (2015). A review of factors influencing the success of community renewable energy mini-grids in developing countries. *Proceedings of the Asia Pacific Solar Research Conference 2015*, Brisbane, Australia.

Manetsgruber, D., Wagemann, B., Kondev, B., & Dziergwa, K. (2015). *Risk management for mini-grids: A new approach to guide mini-grid development*. Alliance for Rural Electrification. Retrieved January 13, 2024, from https://www.ctc-n.org/resources/risk-management-mini-grids-new-approach-guide-mini-grid-deployment

Meliani, M., Barkany, A. E., Abbassi, I. E., Darcherif, A. M., & Mahmoudi, M. (2020). Energy management in the smart grid: State-of-the-art and future trends. *International Journal of Engineering Business Management, 13*, 18479790211032920.

MoWIE. (2013). *Ethiopian national energy policy (2nd draft), Ethiopian ministry of water, irrigation and energy*. Retrieved January 13, 2024, from https://www.lse.ac.uk/GranthamInstitute/wp-content/uploads/laws/1195b.pdf

MoWIE. (2019). *National electrification program 2.0: Integrated planning for universal access*. Technical report, Ethiopian Ministry of Water, Irrigation and Electricity.

Murphy, P. (2008). *Plan C: Community survival strategies for peak oil and climate change*. New Society Publishers.

Nganga, J., Wohlert, M., Woods, M., Becker-Birk, C., Jackson, S., & Rickerson, W. (2013). *Powering Africa through feed-in tariffs. Advancing renewable energy to meet the continents electricity needs*. World Future Council, the Heinrich Böll Stiftung, and Friends of the Earth England. Retrieved January 13, 2024, from https://www.ctc-n.org/resources/powering-africa-through-feed-tarriffs-advancing-renewable-energy-meet-continents

Pokhrel, S., Singal, S. K., & Singh, S. N. (2013). Comprehensive study of community managed mini grid. *International Journal of Emerging Technology and Advanced Engineering, 3*(3), 514–520.

Röder, M., Stolz, N., & Thornley, P. (2017). Sweet energy: Bioenergy integration pathways for sugarcane residues. A case study of Nkomazi, District of Mpumalanga, South Africa. *Renewable Energy, 113*, 1302–1310.

Rogers, J., Simmons, E., Convery, I., & Weatherall, A. (2008). Public perceptions of opportunities for community-based renewable energy projects. *Energy Policy, 36*, 4217–4226.

Romero-Rubio, C., & Díaz, J. R. (2015). Sustainable energy communities: A study contrasting Spain and Germany. *Energy Policy, 85*, 397–409.

SEforALL. (2020). *State of the global mini-grid market report 2020*. SEforALL and BloombergNEF. https://www.seforall.org/publications/state-of-the-global-mini-grids-market-report-2020

Tenenbaum, B., Greacen, C., & Vaghela, D. (2018). *Mini grids and the arrival of the main grid: Lessons from Camodia, Sri Lanka and Indonesia*. Energy Sector Management Assistance Program (ESMAP) Technical Report 013/18. Retrieved January 13, 2024, from https://openknowledge.worldbank.org/handle/10986/29018

Thellufsen, J. Z., & Lund, H. (2016). Roles of local and national energy systems in the integration of renewable energy. *Applied Energy, 183*, 419–429.

USAID. (2018). *Challenges and needs in financing mini-grids*. Retrieved January 13, 2024, from https://www.usaid.gov/energy/mini-grids/financing

Vargas, M., & Davis, G. (2016). *World energy scenarios 2016—The Grand Transition*. World Energy Council.

Walker, G., Devine-Wright, P., Hunter, S., High, H., & Evans, B. (2010). Trust and community: Exploring the meanings, contexts and dynamics of community renewable energy. *Energy Policy, 38*, 2655–2663.

Wirth, S. (2014). Communities matter: Institutional preconditions for community renewable energy. *Energy Policy, 70*, 236–246.

Zebra, E. I. C., van der Windt, H. J., Nhumaio, G., & Faaij, A. P. (2021). A review of hybrid renewable energy systems in mini-grids for off-grid electrification in developing countries. *Renewable and Sustainable Energy Reviews, 144*, 111036.

CHAPTER 8

Energy Literacy for the Energy Transition: Forming the Next Generation of Energy Practitioners in Ethiopia

Getachew Bekele, Adugnaw Lake, Dawit Habtu, and Amare Assefa

Abstract Delivering the energy transition depends on multiple actions in policy and regulation, project delivery, and project operation that require multiple skills. To what extent is current energy education serving the demands of a rapidly changing sector? This chapter adopts the energy literacy framework to examine the energy education landscape in Ethiopia and its suitability to the current demands of the transition to sustainable

G. Bekele (✉) · A. Lake · D. Habtu · A. Assefa
Adis Ababa University, Addis Ababa, Ethiopia
e-mail: getachew.bekele@aait.edu.et

A. Lake
e-mail: adugnaw.lake@aait.edu.et

D. Habtu
e-mail: dawit.habtu@aait.edu.et

A. Assefa
e-mail: amare.assefa@aait.edu.et

V. Castán Broto (ed.), *Community Energy and Sustainable Energy Transitions*, https://doi.org/10.1007/978-3-031-57938-7_8

energy. The energy literacy framework highlights that while energy education has, in general, focused on developing technical and economic skills for the management of energy systems, a changing landscape also requires community and political literacy.

Empirical research conducted in Ethiopia examined first policymakers' perspectives on the skills needed to deliver the energy transition. Second, the research analysed energy-related programmes offered in higher education, looking into the historical development of the Institutes of Technology in Ethiopia. Third, the research conducted a survey with higher education leavers to examine how they have adapted their education to their professional practice. The research reveals a significant gap in social sciences perspectives on energy, compounded by educational programmes that prioritise theory over practice. A concerted reform within the educational sector is needed to ensure professionals acquire the skills to lead the energy transition in Ethiopia.

Keywords Energy literacy · Ethiopia · Institutes of technology · Higher education · Skills for the energy transition

8.1 Introduction

There is an urgent need for a transition to sustainable energy in Africa to provide access to a growing population and to support thriving economies across the continent. The recently published Just Transition expert report (Sokona, 2023) argues that overcoming the current economic challenges in Africa depends on achieving an energy transition and a consistent response to climate change across the continent. They say that "*Africa must scale up energy production and access while leapfrogging dirty-energy systems to modern, affordable, renewable energy systems*" (p. 5).

The transition to sustainable energy comes together with economic promises, including the possibility of providing universal energy access, halting deforestation challenges related to the use of biofuels, and developing an innovative renewable industry that can provide green jobs (Tănasie et al., 2022). There is a pressing demand for renewable energy specialists with the knowledge and skills to design, instal, and maintain renewable energy systems. The need for renewable energy education and training is ever more present worldwide. Education is crucial in transforming human behaviour to improve energy use and represents a long-term investment in addressing energy-related challenges (Gebremekel et al., 2019). The question is, however, the extent to which current education enables bridging energy delivery with the needs of the communities to require energy access, seeking to build a people-centred energy sector that delivers for the development of the well-being needs of people in different African countries.

In Ethiopia, the energy transition offers multiple opportunities to facilitate economic recovery post-war. While the civil war in Tigray represents a setback for the country's advancement towards development outcomes, the energy transition has the potential to accelerate recovery. However, one challenge to delivering the energy transition is the development of appropriate skills to participate in the benefits of the green economy and envisage renewable futures at the national, regional, and local levels. This chapter interrogates whether Ethiopia's current educational context is equipped to provide people with those skills.

This chapter starts from the normative point of view that the requirements of energy literacy require diverse skills at different stages, from policymaking to planning, from the construction and delivery of projects to their management. This means there is a need to train people for different jobs at all stages of development. In Ethiopia, engineering education has been the leading provider of energy workers. However, the demands of the energy transition are such that a wide range of professionals are needed. This is a question that requires multi-disciplinary insights. For example, financial managers are required to manage energy companies. Political scientists and social scientists are needed to develop policy. Community managers and sociologists are needed to manage engagement with a range of communities whose needs may need to be understood. Social psychologists may be needed to understand the processes of behaviour change and demand management. Artists have proven to be able mediators of community engagement. In sum, engineering alone cannot provide complete responses to the challenges of the energy transition.

At the same time, the energy transition depends on a range of professionals who enter the sector through professional practice, only sometimes requiring a university education. From jobs in construction to day-to-day management, the green energy sector offers a wide range of opportunities to professionals at all levels. Many of these professionals, however, do not have access to skills development programmes. Finally, the energy transition requires the development of forms of energy literacy for everyone in society as people develop increasing levels of autonomy to engage with the energy sector.

One of the challenges for the green economy has been the dominance of masculinist ideas in the workforce (Clancy & Feenstra, 2019). Women are routinely excluded from the energy sector and not always considered active agents of the energy transition (Pueyo et al., 2017). This is seen in their access to energy engineering education and their absence from the energy industries. Such a lack of attention to social diversity limits the possibilities for an energy transition in many African countries.

What are the energy literacy skills needed for the energy transition in Ethiopia, and how are they met? First, our analysis examines policymakers' perspectives on the transition and the kind of skills that will be able to meet those demands. Next, the chapter focuses on the field of energy education, interrogating its gaps for the future through a systematic examination of energy engineering programmes in Ethiopia. Third, the chapter examines the delivery of engineering programmes: who are the graduates of those programmes, and what do they do? The analysis suggests that the energy transition requires a wide range of skills and people, but this is not entirely understood from policymakers' perspectives. This research shows that the current workforce is masculinized, and women tend to be excluded from the energy professions. There is also limited lifelong learning across the sector, which prevents innovation from developing. Finally, existing educational programmes are exclusively technical. In summary, there is an overall focus on engineering that limits broader approaches to the energy transition, especially innovation. The chapter concludes with a reflection on the need to breach the skills gaps of professionals within the energy sector for a transition to sustainable energy.

8.2 Energy Education and Sustainable Energy Transitions

Energy education empowers people to address questions and tackle problems related to energy. The Energy Literacy Framework outlines several "Essential Principles" and a collection of Fundamental Concepts that align with each principle (Brown et al., 2015). Instead of encompassing every aspect of energy understanding, the framework emphasises the fundamental concepts essential for all individuals to make informed energy decisions and engage in public discussions about energy (Brown et al., 2015). The objective for individuals is to develop a solid foundation of energy literacy, enabling them to navigate energy-related topics with greater understanding and engage in informed conversations about energy policy, technology, and sustainability. The framework provides a structured approach to gain the necessary knowledge and skills to make informed decisions regarding energy production, consumption, and conservation. It also promotes critical thinking and empowers individuals to evaluate various energy sources and systems' environmental, economic, and social impacts. The Energy Literacy Framework is vital in fostering energy education by providing a structured and comprehensive set of principles and concepts that underpin energy understanding for all citizens. It has, however, not been widely adopted across many energy education programmes, many of which remain wedded to fundamentally technical programmes that do not always recognise the wide range of factors that influence an energy transition.

The Energy Literacy Framework focuses on expanding the range of subjects in energy education, essential for delivering the energy transition (see Fig. 8.1). For example, incorporating renewable energy as a subject of study in energy-related topics can serve as a novel way to engage and motivate students, particularly those who are conscious of environmental issues. By exploring the analysis and classification of renewable energy sources and understanding their origins, students develop a deeper understanding of renewable energy and its significance and have the basis to facilitate energy innovation. Energy conservation has also gained prominence in energy educational programmes.

Energy-educated individuals play a crucial role in achieving universal access to electricity through various means, including expertise in the renewable energy (RE) supply chain, involvement in policymaking and regulation, and active participation in the larger community. Becoming experts in the RE technologies supply chain is essential for community energy to contribute to universal access to electricity effectively. Assessing the capacity for energy learning helps determine if sufficient skills are available to support community energy projects, as the sector demands growing numbers of technicians. By equipping graduates with technical skills, energy education addresses the existing skill and knowledge gap in the RE supply chain, covering everything from the procurement of

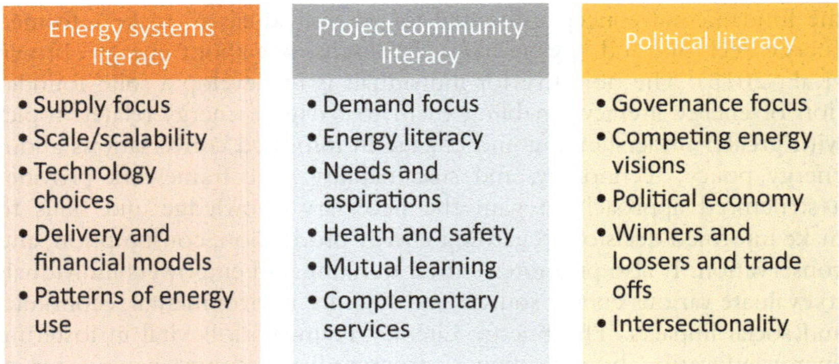

Fig. 8.1 Social systems literacy (adapted from Cloke et al., 2017)

renewable energy technologies and accessories to their installation and maintenance in off-grid solutions.

Different types of community energy, such as micro-hydropower, solar photovoltaic, solar steam generators, biogas, and wind-power generator technologies, require specific skills and occupations associated with each source's supply chain. The RE value chain encompasses various stages, including equipment manufacturing and distribution, the development of renewable energy projects, construction and installation, operation and maintenance of renewable energy facilities, and cross-cutting activities contributing to multiple stages. Bioenergy has an additional step involving the growth and harvesting of biomass. Most of these stages require specialised skills. However, there is also a need for integrative skills across the different stages of supply. Energy graduates also fill gaps in policymaking, joining government institutions such as the Ministry of Water, Irrigation, and Energy (MoWIE), Ethiopian Electric Power (EEP), Ethiopian Electric Utility (EEU), and Ethiopian Energy Authority (EEA) in the case of Ethiopia.

Locally specific knowledge grounded on communities is also crucial for achieving the goal of universal access to electricity. The level of community knowledge on sustainable development significantly impacts the outcome. Within communities, individuals who understand energy are more likely to prioritise and engage in energy transitions towards renewables. Energy-educated individuals within the community can make informed decisions and actively participate in low-carbon energy transitions, mainly through off-grid solutions. Assessing energy learning capacity helps determine how the community supports community energy as an electrification method. Highly educated professionals can play a crucial role in mediating and integrating community knowledge into transition processes.

8.3 Key Priorities in the Energy Transition in Ethiopia

Over the past three years, Africa's population has consistently expanded at an average annual rate of 2.5%, surpassing all other regions and doubling the global average. Only four countries—the Democratic Republic of the Congo (DRC), Egypt, Ethiopia, and Nigeria—account for almost 40% of Africa's population. Projections indicate that Africa's population is set to experience substantial growth, estimated to surpass 1.7 billion by 2030. The demand for electricity in Africa is projected to increase by approximately 75% from 680 TWh to 1,180 TWh by 2030. The surge in energy consumption in Africa primarily stems from the combined effects of economic and population expansions. The Africa Energy Outlook 2022 (IEA, 2022) indicates that the final consumption of modern fuels in Africa, about end-uses, experiences an average yearly growth of 5% between 2020 and 2030 in the sustainable Africa scenario (SAS). This marks a noteworthy increase compared to the 2% growth observed in the previous decade. Furthermore, the modern primary energy supply exhibits an average annual growth rate of 3% from 2020 to 2030. However, it is worth noting that the total primary energy supply, including the traditional utilisation of solid biomass, is projected to decline by 13% by 2030.

In Ethiopia, the energy transition is an urgent demand. Ethiopia has connected 33% of its population with on-grid and 11% with off-grid solutions—mostly mini-grids and solar PV systems—totalling 44%. In addition, 90% of the population still relies on biomass for cooking. The government's goal is to achieve universal electricity access nationwide by 2025, according to the National Electrification Program (NEP) titled "Light to All" (Gebremekel et al., 2019). The NEP focuses on implementing a cost-effective grid connection strategy, considering the geographic distribution of households to connect 65% of the population to the grid. The remaining 35% will receive an off-grid power supply.

The off-grid power supply is specifically designed to cater to remote settlements and villages where grid connectivity is not the most viable option. The intended technologies for this purpose primarily include standalone solar systems supplemented by mini/micro grid network connections, which are expected to be facilitated by the private sector. However, considering the current level of accessibility, it appears that achieving this ambitious goal within the stipulated timeframe may present significant challenges and may not be readily attainable.

Community energy (CE) projects aim to enable citizens to own or participate in generating sustainable energy. This can be achieved if citizens (private households, communities, etc.) collectively form a legal structure to finance and establish such projects (See Chapter 1). The electricity generated by such projects is then consumed locally or collectively sold to local power utilities, and profits can be split among participating citizens or re-invested in other projects. Citizens may establish, develop, and own projects by themselves (bottom-up) or involve other actors, such as energy utilities, to develop the projects in which the community can participate (top-down). This latter modality may have greater viability in Ethiopia, given the government's ferreous control over the energy sector. In any case, community energy is still a rarity in Ethiopia.

8.4 Policymakers' Perspectives on a People-Centred Energy Transition

A survey was conducted to collect the opinions and views of various representatives of the key stakeholders in the energy sector with roles of policy development, regulation, and power supply in the whole process of creation and implementation of community energy in Ethiopia, seeking to understand the main priorities for the transition to sustainable energy in Ethiopia within the existing legislation, the multiple ways in which communities are involved in the delivery of sustainable energy, and the training and education needs arising in this context. The survey questions were electronically distributed to 26 representatives of stakeholders with different occupations, including high-level executives, policymakers, directors, senior experts, and lecturers.

The survey shows that for the central government, building technical capacity for the transition is a crucial priority, alongside providing finance and strengthening the regulatory framework. Higher education institutions also understand that they have a role to play in the transition through their regular education and by providing long-life training to local technicians who can operate and maintain the system alongside the formation of engineers.

The survey also shows that there is a significant awareness gap about the general idea and use of community energy or energy communities. Still, there is not a social perspective on the delivery of the energy transition. There is legislation (Proclamation No. 317/2003) that promotes cooperative engagement in rural electrification activities through loan-based finance, but it doesn't clearly and specifically deal with "community energy". The results have also shown that the concept of community energy is relatively new to the country and not entirely explored, even though MoWIE and GIZ made some initiatives with the development of micro-hydropower mini-grids through cooperative modality.

Respondents agreed that involving communities to address electricity access challenges could be feasible and effective. Regarding the feasibility of community energy projects, most respondents stated that such projects could be feasible in Ethiopia with the right policy, legislation, and incentives in place. However, several interviewees raised limitations in finance, technical skills, management, and governance. Several interviewees highlighted that solar-based generation, including solar mini-grids, is highly attractive to communities in Ethiopia. Still, they unanimously argued that technology selection is site and resource-dependent and that, rather than making prescriptions, technology decisions depend on detailed site studies.

As explained above, all interviewees anticipated challenges in developing and implementing community energy projects in Ethiopia. The main ones include limited awareness about community energy at all levels of decision-making, lack of clear and specific policies, regulatory guidelines and incentives, and financial constraints, especially lack of access to finance for communities. Interviewees also expressed doubts about the availability of technical skills to run community energy projects, disruptions in the supply chain, which may lack renewable components, an underdeveloped renewable industry, and limited public acceptance of projects of such dimensions.

Measures were proposed to overcome these challenges and create an enabling environment for community energy projects. Most interviewees argued that the development of community projects, and off-grid energy more generally, requires the government to take the initiative by creating awareness, setting clearly outlined policies, incentives, and regulations, establishing strong institutions that can provide all-rounded support and coordinate all stakeholders, setting legal frameworks and finance in place (especially arranging various finance accesses to overcome the financial problems of the community), encouraging local manufacturing of renewable systems and components, and starting pilot projects in selected villages. In doing so, the stakeholders' perspectives align with existing energy policy and the assumption of government dominance.

Thus, other actors such as power utilities, finance institutions, NGOs, universities, and private companies are given a secondary role to work in collaboration and closely engage in public awareness creation, policy advice, and consultation, provide training, share the experiences of other countries which implemented similar projects, make research on detail implementation, provision of funds, support hard currency financing for import of system components and spare parts, enhance private sector involvement, engage local leaders and work closely with local governments and administrators. Central to the development of this ecosystem of skills is a thriving higher education servicing of the energy sector, including appropriate programmes to deliver technical, managerial, and social engagement skills, putting energy education at the heart of the transition to sustainable energy in Ethiopia.

Stakeholder perspectives are thus aligned and support the government's vision for an energy transition, structuring agencies and programmes of action and, thus, framing off-grid energy (especially community energy) as an addition to the main electrification efforts. This is an engineering-led transition that may overlook the complex social impacts of such a process of change. One way to challenge this monolithic transition narrative is, precisely, through education and the cultivation of a wide range of energy literacy skills. This also entails recognising the wide range of skills needed for a transition beyond higher education, and the interdisciplinary nature of the challenge.

8.5 The Context of Higher Education for Energy in Ethiopia

The United Nations Educational, Scientific and Cultural Organization (UNESCO) has argued that higher education shapes any country's socio-economic and political development (UNESCO, 1998). A pivotal moment in Ethiopian history was in December 1950 when Emperor Hailesellassie I University College (currently Addis Ababa University) first admitted 80 students from three secondary schools for higher education studies (Habte et al., 1963).

In 1994, the Federal Democratic Republic Government of Ethiopia (FDRE) ratified a Higher Education Proclamation (HEP) to lay down the foundation of a legal framework to enable higher education institutes to engage in research on problem-solving topics to utilise national resources and provide academic freedom to expand higher education in the nation (FDRE, 1994). More than 55 public and five private universities are operational, with 306 registered private higher education institutions. By the end of the 2019/2020 academic year, 212 undergraduate, 457 masters, and 220 PhD programmes had been imparted at all public universities (Tegegn, 2021).

Engineering education, both undergraduate and postgraduate, has been monopolised by public universities ever since higher education started. There are 10 Institutes of Technology (IOT) and universities with engineering colleges fully engaged in engineering education. Energy technology and energy-related courses are taught within those universities in undergraduate and postgraduate programmes. Engineering and technology programmes take a 32% share of enrolment from all undergraduate programmes throughout the universities (MOE, 2017). The higher education reform was implemented in 2009 on the admission of students from secondary school with a 70:30 quota scheme whereby 70% of undergraduate students study sciences and engineering/technology and 30% for social sciences, assuming hard science and engineering/technology graduates contribute to country development. This reform was again revised in 2019 to be replaced by a 55:45 placement for hard science and social science, respectively (Yirga, 2020). Although there is no dedicated energy-related BSc curriculum in public universities, there are different master's programmes in Energy technology and sustainable energy engineering at a few universities nationwide.

For example, the Addis Ababa Institute of Technology, formerly known as the Imperial College of Engineering, is the oldest institution teaching engineering education and is considered the leading Institute of Technology in Ethiopia. It was established in 1953 and initially offered two-year intermediate engineering programmes. Over the years, the college expanded its programmes, introduced four-year degree programmes in civil and industrial engineering, and later split the industrial engineering programme into electrical and mechanical engineering.

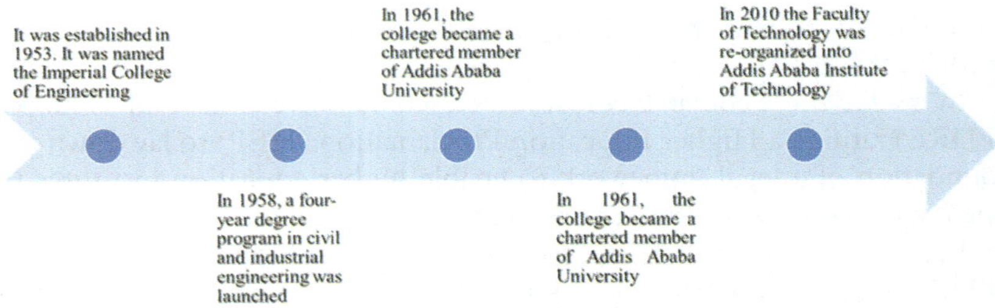

Fig. 8.2 Key events in the evolution of Addis Ababa University

In 1961, the college became a chartered member of Addis Ababa University. In April 2010, the Faculty of Technology was reorganised and given autonomy, leading to the establishment of the Addis Ababa Institute of Technology. The institute has a significant student population of over 5500 undergraduates and 4500 graduate (MSc and PhD) students. It also boasts a highly qualified staff of 400 academic members and around 600 administrative and support staff (MOE, 2018). A strong commitment to energy engineering has shaped the higher education landscape (Fig. 8.2).

8.6 Engineering Higher Education in Ethiopia and the Energy Transition

The engineering sciences dominate thinking about energy in Ethiopia. How is engineering education conceived and delivered? To answer that, the Addis Ababa team developed methods to assess the presence, nature, and geographical distribution of energy programmes and courses offered by Ethiopian public universities. The following steps were taken:

1. Compilation of University Programs: A thorough compilation of Ethiopian public universities and their respective programmes was conducted using the official website of the Ministry of Science and Higher Education (MoSHE), to identify universities offering energy programmes and courses.
2. Selection of Energy Programs: Energy programmes and courses at the BSc, MSc, and PhD levels were targeted for analysis. The analysis went systematically through the list of universities to identify energy-related programmes, courses, and modules.

3. Curriculum Review: A systematic curriculum review was conducted for each identified university. This involved a detailed examination of the course structure, descriptions, and learning outcomes of energy programmes and courses. The focus was on understanding the subject matter, level of specialisation, and overall content of the curriculum to gain insights into educational offerings in the energy domain.
4. Survey on Existing Educational Programs: To gather additional information about the energy programmes and curricula, a survey questionnaire was prepared using Google Forms. The questionnaire was sent to the School/Department Heads or Stream Chairs of the Institute of Technologies (IoTs) and Engineering Universities. The survey aimed to collect data on the existing educational curriculum of public universities, specifically focusing on the programmes of interest.
5. Data Analysis: The analysis focused on determining the percentage of universities offering energy programmes and courses at the BSc, MSc, and PhD levels. It provided insights into the prevalence of energy education across different academic levels and helped assess the overall landscape of energy learning in Ethiopian public universities.
6. Mapping Energy Learning: The geographical distribution of universities offering energy programmes and courses was visualised using Public Tableau. This mapping exercise facilitated a better understanding of the regional availability and accessibility of energy education. The results of the mapping exercise were shared and made publicly accessible through the CESET (Community Energy Solutions and Empowerment Technologies) website, enabling the dissemination of information and promoting awareness about the geographical distribution of energy education in Ethiopian public universities.

By combining the compilation of university programmes, curriculum review, survey responses, and mapping exercises, this comprehensive methodology provided a holistic assessment of the energy programmes and courses offered by Ethiopian public universities. It enabled insights into the presence, nature, and regional distribution of energy education, serving as a valuable resource for policymakers, educators, and stakeholders in the field.

The analysis shows significant capacity for the development of technical skills in Ethiopian public universities but a lack of other complementary skills to understand energy systems. The initial inventory shows that 45 public universities and ten institutes of technology (IoTs) deliver energy education in Ethiopia. The proportion of IoTs in the country is shown in Fig. 8.3, indicating that only 22% of the universities in Ethiopia have IoTs.[1]

Across these institutions, energy education is offered in various ways. Some universities provide standalone MSc and/or PhD programmes in energy. In contrast, others incorporate energy courses into existing programmes such as Electrical Power Engineering, Electrical Power Systems Engineering, and Thermal Engineering. Figure 8.3 outlines the titles of programmes offered at different universities (Fig. 8.4).

Figure 8.4 shows the frequency of those programmes. There is no dedicated energy programme at the undergraduate level. However, two energy modules are being taught at different IoTs and engineering universities, primarily targeting electrical power engineering students. These courses are "Energy Conversion and Rural Electrification" and "Hydropower Engineering".

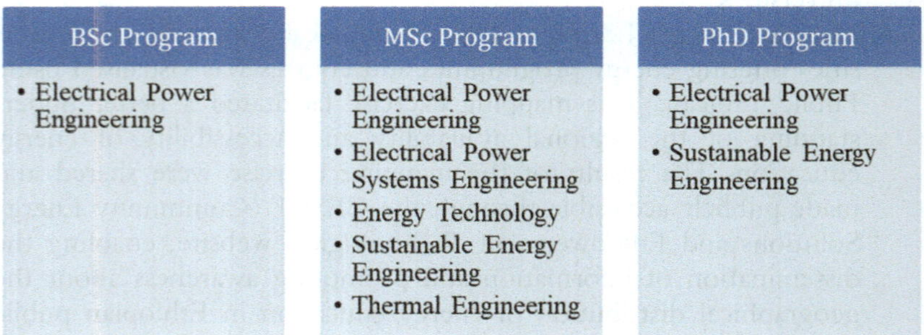

Fig. 8.3 Titles of energy-related programmes offered at different IoTs in Ethiopia

[1] The list of IoTs are: Addis Ababa Institute of Technology, Arba Minch Institute of Technology, Bahir Dar Institute of Technology, Dire Dawa Institute of Technology, Ethiopian Institute of Technology—Mekelle, Haramaya Institute of Technology, Hawassa Institute of Technology, Jimma Institute of Technology, Kombolcha Institute of Technology, and Mekelle Institute of Technology.

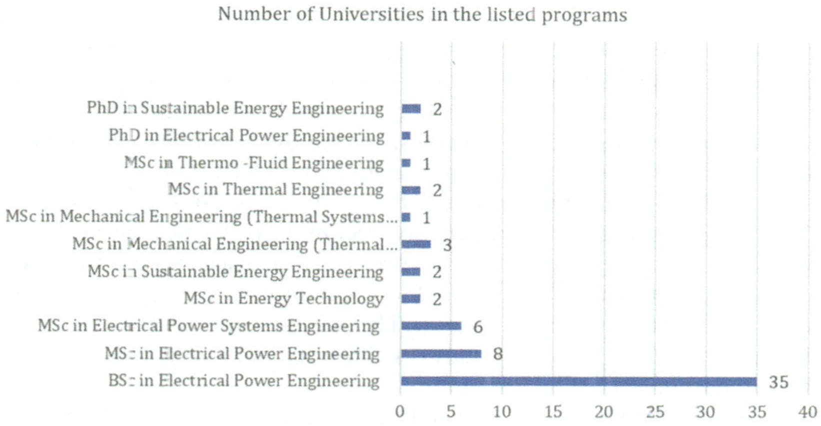

Fig. 8.4 The number of universities providing energy education

The module "Energy Conversion and Rural Electrification" aims to introduce students to various primary energy resources and associated technologies for converting them into heat and electricity. The course covers renewable and non-renewable technologies, specifically focusing on hydropower, solar, wind, geothermal, and biomass resources. It provides an overview of conventional and non-conventional power plants, insights into planning and designing small-scale and off-grid electrical power systems, and techniques for rural electrification planning and design. This course is the one that offers the most holistic approach to energy education.

The module "Hydropower Engineering" focuses on the working principles and major components of hydropower plants, covering everything from the catchment area to the turbine generator and tail race. The course aims to provide insights into hydropower engineering concepts, knowledge of planning, designing, and developing hydroelectric power plants, understanding the design of dams and spillways, and familiarisation with hydraulic turbine operation. These two courses are offered at the BSc level in 34 engineering universities, representing 76% of the public universities in Ethiopia (the remaining 24% do not offer engineering education).

At the MSc level, there are energy programmes available in four IoTs: Addis Ababa Institute of Technology (AAiT), Ethiopian Institute of Technology—Mekelle (EITM), Bahir Dar Institute of Technology (BDIT), and Jimma Institute of Technology (JiT). These programmes cover a wide range of energy-related courses, including quantitative methods for energy studies, renewable energy resources and technologies, energy economics and policy issues, bioenergy systems engineering, wind energy systems, fuel processing technologies, hydropower systems engineering, photovoltaic systems engineering, energy project planning and management, energy conservation and the environment, solar thermal systems engineering, and geothermal energy technology. While the technical disciplines are represented in all these courses, the social sciences are not.

In addition to the IoTs, other MSc programmes in different universities also offer energy-related courses. The specific courses vary depending on the programme and university, as illustrated in the energy learning capacity dashboard.[2]

The results indicate that energy learning capacity at the MSc level is more comprehensive compared to the BSc level, but in both cases, there is a significant absence of social sciences skills in educational programmes and near-zero consideration of training of technicians and associated professions to work in the energy industry.

In conclusion, the investigation into energy technology programmes and courses in Ethiopian universities reveals that while energy education is not offered as a standalone programme at the undergraduate level, there are energy courses in engineering programmes across the country. The courses "Energy Conversion and Rural Electrification" and "Hydropower Engineering" provide students with knowledge of different energy resources and associated technologies, including conventional and non-conventional power plants, renewable energy sources, and planning and design of electrical power systems. However, the investigation reveals a dearth of social and policy skills embedded in the university programmes, limited engagement with innovation, and a lack of attention to the intermediary technical skills on which the transition depends. Energy education needs to foster a culture of creativity, innovation, and entrepreneurship by drawing on a wide range of skills and providing access to energy education to an increasingly diverse workforce. Meanwhile, the existing programmes match the general expectations of policymakers and respond to the current needs of the energy system. The challenge now, however, is the perspective of an energy transition in Ethiopia, for

[2] https://cesetproject.com/higher-education-programmes-teach-energy-related-skills-ethiopia, retrieved January 16, 2024.

which hydropower may not suffice. The analysis suggests that some essential skills and forms of knowledge crucial for the transition, emphasised in the energy literacy framework, are missing from current educational programmes. Universities and policymakers could make a difference in the transition by revising their offer in relation to expected, rather than present, needs.

8.7 The Perspectives of Energy Technology Graduates on the Energy Transition

In addition to the perspectives on education from policymakers and the existing educational programmes, an additional question was understanding the experiences of graduates who then join the energy sector. Who are these graduates, and what are they doing? This research took a purely quantitative approach, focusing on technology graduates from IoTs with some experience in the job market. The questionnaires, in a Google Spreadsheet, were distributed via email and were returned by 26 graduates, only four of them women. In-depth interviews were conducted either in person or on online platforms like Zoom and Google Meet with 16 graduates, four of whom were women. In-depth interviews followed a conversation about post-graduation experiences, ranging from 24 to 85 minutes.

In the past two and half decades, higher education admissions have increased in line with the expansion of new government universities across the country. The Ethiopian government's University Capacity Building Program (UCBP) was established to increase the infrastructure sector in higher education to generate a high number of engineering and hard science graduates to enhance the growth and transformation plan of the country (Alemayehu, 2021). The number of admitted regular undergraduate students enrolled in public universities has increased dramatically from 98,404 in 2014 to 388,186 students in 2018. Furthermore, the workforce, undergraduate regular students' graduation rate boosted significantly from 7,006 in 2004 to 62,199 in 2018 (Mekonnen Yimer et al., 2022).

Ethiopia's higher education gender mix has a significant variation among female and male undergraduate students across the country. In Ethiopia's case, in the 2018/19 academic year, in both public and private undergraduate students, the ratio was 63/37 male to female (MOE, 2018). Global data from 74 countries between 1995 and 2018 showed that the male ratio in higher education enrolment increased from 0.95 to 1.14 on average, whereas sub-Saharan African countries have among the lowest global female-to-male enrolment ratios, about 0.6 (UNESCO, 2021). In Ethiopia, female enrolment in public and private universities in hard science and engineering is among the lowest with a 0.59 value of female-to-male ratio (MOE, 2018).

The background of the study group was evaluated in terms of gender, B.Sc. qualification, year of B.Sc. graduation, and years of experience based on the information collected from the responses to the questionnaire (Fig. 8.5). Energy-related postgraduate programme admission requires a first-degree in Electrical Engineering, Mechanical Engineering, Agricultural Engineering, and other hard sciences based on the admission criteria of the respective university. In our survey, most students are mechanical engineering graduates (Fig. 8.5). This is a relatively small sample, but it also suggests that women are underrepresented in the sector (Fig. 8.5).

Sixty-two per cent of respondents are from different regional university staff members who came with scholarships from the government to study renewable energy-related programmes to enhance their respective departments, while the other 38% of respondents are from the public and private sectors engaged in the engineering field.

Among the respondents, 71% of graduates from energy-related master's programmes are currently employed in universities as lecturers. The others are working in positions that are not relevant to their specialisation, and only 4% of them are working in the fields directly related to their study. Individuals in this sample have joined the second-degree

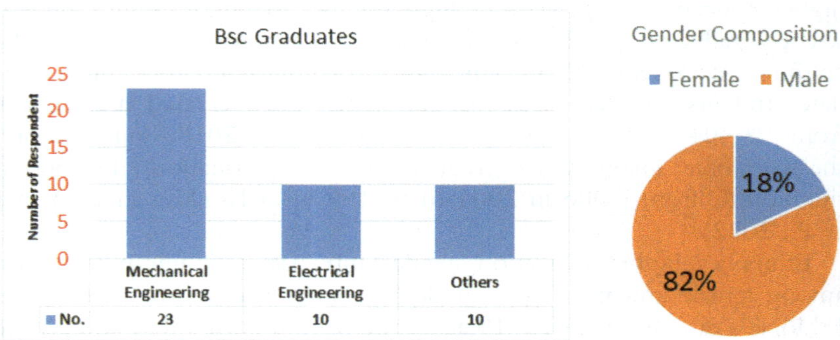

Fig. 8.5 Background and gender of participants in the questionnaire

specialisation in energy-related programmes from different industry areas, which include private consultants, researchers, and higher institution instructors. The length of work experience after first-degree graduation varied from 1–5 years, 6–10 years, and greater than 10 years, accounting for 42.9%, 35.7%, and 21.4%, respectively.

Among the respondents, 53.6% did not ever have a chance to meet the rural population who need renewable energy technologies, whereas 46.4% had a chance to visit the rural population in need. They have pointed out that lack of awareness, lack of finance, lack of infrastructure, and substandard materials in the market are taking the lion's share for having a low community-based energy system in the rural population. From the data collected, postgraduate energy-related education is not easily available in all IoTs and universities across the country. Energy graduates will play a vital role in alleviating the rural population's energy issues if the current education landscape is changed to accommodate more energy-related postgraduate programmes introduced in the country.

According to the interviews, short-term training in the engineering sector has a significant impact on the performance of graduates in their specific fields. Training might be organised in person or online based on the contents of the short-term training. Among the respondents, only 39% of respondents experienced short-term training (online or in person). Topics focus on solar energy development, small-scale hydropower development, wind energy, and biomass. Renewable energy simulation and software training are among the training taken by a few respondents.

The master's programme is less prestigious than master-level programmes in other engineering programmes. About 92% of respondents consider the majority of master's degree level courses to be delivered as business as usual, in which the professors lecture using a projector supported by corresponding examples and written assignments. Oral presentation and research-based assignments are used very rarely in all IoTs. According to the respondents, laboratory-supported courses are not delivered intensively in all IoTs where the programme is available. A few courses like bioenergy engineering, thermal engineering, and wind energy have laboratory exercise demonstrations. The IoTs do not have enough laboratory facilities dedicated to postgraduate students. Courses like PV engineering, wind engineering, solar thermal engineering, biofuel, and hydropower engineering are among the major courses proposed by responders to have dedicated laboratory exercises in parallel with the courses. They have observed that the courses are not practice-oriented and are not supported by rigorous simulation software. They have recommended that professors should involve graduate students in scientific research- projects with simulation software. Respondents argue that undergraduate mechanical and electrical engineering courses should be revised to provide wider coverage of renewable energy courses. Mechanical graduates proposed basic electrical engineering courses, and renewable energy-based courses that are more relevant to support energy-related postgraduate courses. For electrical engineering graduates, courses in mechanical engineering like heat transfer, thermal engineering, and electrodynamics can support postgraduate courses. In sum, there is a clear lack of practice-oriented training and short-term courses to develop the specialist skills that would be needed in a transition to sustainable energy.

Interviewees explain that a major issue with all IoTs is an acute shortage of professors who have specialised in renewable and energy-related topics to provide lectures. All IoTs share a staff member as a guest professor. The lowest academic rank teaching postgraduate courses is a lecturer with a master's degree in most IoTs. Twenty-two per cent of respondents were taught by an assistant professor with a master's degree as the lowest academic rank. In comparison, 48.1% were taught by lecturers with a master's degree as the lowest academic rank, and 29.6% were taught by an assistant professor with a PhD degree as the lowest academic rank. This figure shows that the master's programme in energy-related courses is open on IoT without enough academic staff with PhD degrees. This means that lecturers may face strenuous challenges in delivering these programmes in the time allocated, with further ramifications for the well-being of lecturers and for students' learning outcomes.

8.8 Conclusion

Delivering transitions to sustainable energy in African countries will require substantial skills, both technical and social, to manage supply chains, change regulations, and engage communities on one of the major challenges of this generation. Development of energy literacy requires engaging a wide range of audiences in energy education, from the regulators in charge of energy frameworks and the investors to the people who will benefit from the supply of energy. As community energy becomes a workable alternative to deliver energy transitions in African countries, the need to understand communities and their role in delivering energy projects becomes more urgent. Energy education needs to engage with such challenges that enable an innovative and just vision of the future of energy in the continent.

This is particularly salient in Ethiopia, where the recent civil war has put into question many of the recent gains made in terms of energy access across the country. The war has destroyed not only infrastructure but also capacities and skills. It has also affected teaching programmes all across universities. Higher education can help build a foundation to deliver a renewable energy future and bring renewed prosperity across the country.

This research, mostly conducted before the war started (which would have exacerbated the challenges), shows that the current landscape of energy education in Ethiopia needs to evolve to provide workable responses to the energy transition. First, there is a challenge in the perceptions of a transition to sustainable energy, which shapes existing policies, cultures, and actions. Key stakeholders paint a technologically led future with big renewable infrastructures, mainly hydro, without recognising the diversity of actions that will enable engaging the population in a broader cultural change. Second, there is a challenge in the structure of energy-related programmes provided at public universities, especially IoTs. They have an exclusive technical focus with limited engagement with issues of rural development and only in some undergraduate programmes. System-level questions about the politics and governance of energy resources and infrastructures, as well as sociological and socio-psychological questions related to the processes of technological and behavioural change during the transition, should be considered. The just transition is not something studied in these programmes. When examining this against the background of the energy literacy framework, there are substantial gaps in the energy studies curriculums in public universities in Ethiopia. There needs to be a holistic understanding of a wider range of issues affecting the transition, including political debates, policymaking and implementation, community engagement, and behaviour change. Moreover, there appears to be only limited interest in training technicians and lifelong learning, which may be key to ensuring continuous innovation within the energy sector. Third, the experiences of graduates show that there is a need to develop more engaged education programmes actively engaged with practice. While budgetary issues may pester public universities, a practice-oriented education around student-led projects and laboratories may be cost-effective and reduce the need for contact time with already stressed lecturers.

An important finding of this research is that the already significant gender gap in higher education in Ethiopia is likely to be even higher in the energy sector. This means that higher education is not supporting the closing of the gender gap that hinders the energy transition. Taking active measures to attract women to energy education through information and awareness campaigns, for example, could be a means to defy the challenges to develop a workforce capable of leading and shaping the transition to renewable energy in Ethiopia.

Investing in energy skills should be a priority for the government and international donors in the post-war period, putting at the centre, the need for diverse skills and diverse people shaping the future of energy in Ethiopia and Africa.

References

AFDB. (2021). *Country strategy paper 2023–2027 and 2022 country portfolio performance review*. Addis Ababa: AFDB

Alemayehu, E. (2021). Expansion of higher education in Ethiopia: Challenges and opportunities: The case of Kotebe Metropolitan University and Saint Mary University. *International Journal of Research in Engineering & Management, 4*(4), 157–179.

Brown, E., Cloke, J., & Harrison, J. (2015). *Governance, decentralisation and energy: A critical review of the key issues*. Loughborough University.

Clancy, J., & Feenstra, M. (2019). *Women, gender equality and the energy transition in the EU*. Publications Office of the European Union.

Cloke, J., Mohr, A., & Brown, E. (2017). Imagining renewable energy: Towards a Social Energy Systems approach to community renewable energy projects in the Global South. *Energy Research & Social Science, 31*, 263–272.

FDRE (1994). *Education and training policy*. Addis Ababa: FDRE.

Gebremekel, D. H., Beyene, G. B., & Ahlgren, O. (2019). *Assessment of resource adequacy in power sector reforms of Ethiopia*. IEEE.

Habte, A., Gebre, M., & Monika, K. (1963). Higher education in Ethiopia. *Journal of Ethiopian Studies, 1*(1), 1–8.

IEA. (2022). *Africa energy outlook 2022*. IEA.

MOE. (2017). *Education statistics annual abstract*. Addis Ababa: MOE.

MOE. (2018). *Education statistics annual abstract*, Addis Ababa: MOE.

Sokona, (2023). *Just Transition*. Retrieved September 28, 2023, from https://justtransitionafrica.org/wp-content/uploads/2023/05/Just-Transition-Africareport-ENG_single-pages.pdf

Pueyo, A., Spratt, S., Bawakyillenuo, S., & Osiolo, H. (2017). *Green investment diagnostics for Africa: what are the binding constraints to investment in renewables in Kenya and Ghana? IDS Research Report 83*. Retrieved January 13, 2024, from https://www.ids.ac.uk/publications/green-investment-diagnostics-for-africa-what-are-the-binding-constraints-to-investment-in-renewables-in-kenya-and-ghana/

Tănasie, A. V., Năstase, L. L., Vochiţa, L. L., Manda, A. M., Boţoteanu, G. I., & Sitnikov, C. S. (2022). Green economy—green jobs in the context of sustainable development. *Sustainability, 14*(8), 4796.

Tegegn, E. (2021). *HERQA starts licensing institutions for online education*. Retrieved September 25, 2022, from https://www.thereporterethiopia.com/10948/

UNESCO. (1998). *World declaration on higher education for the twenty-first century: Vision and action*. UNESCO.

UNESCO. (2021). *Women in higher education: Has the female advantage put an end to gender inequalities?* UNESCO.

Yimer, B. M., Demissie, M. M., Herut, A. H., Bareke, M. L., Agezew, B. H., Dedho, N. H., & Lebeta, M. F. (2022). Higher education enrollment, graduation, and employment trends in Ethiopia: An empirical analysis. *Preprints*. Retrieved January 13, 2024, from https://www.preprints.org/manuscript/202206.0093/v1

Yirga, A. (2020). *Higher education reforms must be informed by evidence*. University World News.

CHAPTER 9

Pioneering Community Energy for Development in Malawi

Christopher Hara, Chrispin Gogoda, and Maxon L. Chitawo

Abstract Community energy projects are increasingly common in Malawi to provide energy access in a country with the lowest electricity access rates in the world. However, they face challenges, some of which only become apparent during the process of implementation. This chapter asks what the lessons from the combined experiences of different implementing community energy projects in different locations in Malawi are.

The research took a comparative approach using primary data from five field visits to community energy sites in the districts Rumphi, Mchinji, Dedza, Mulanje and Nsanje and a literature review including project documents. The analysis shows that community energy projects have a net positive impact on communities, but they face challenges related

C. Hara (✉) · C. Gogoda · M. L. Chitawo
Mzuzu University, Mzuzu, Malawi
e-mail: hara.c@mzuni.ac.mw

C. Gogoda
e-mail: gogoda.c@mzuni.ac.mw

M. L. Chitawo
e-mail: chitawo.m@mzuni.ac.mw

V. Castán Broto (ed.), *Community Energy and Sustainable Energy Transitions*, https://doi.org/10.1007/978-3-031-57938-7_9

to dealing with an inadequate regulatory system, facilitating community participation and leadership and managing unexpected events. The recommendation is to provide long-life and ad hoc support to these projects, for example, via government subsidies.

Keywords Community energy · Community participation · Government subsidies · Malawi

9.1 Introduction

Off-grid mini-grids and other decentralized energy systems can facilitate local electrification, and they have recently been identified as having the potential to facilitate an economic boom in African regions with gaps in electricity access rates (EEP Africa, 2018). They provide communities with the agency to lead such a process of local electrification. However, a common refrain heard in policy circles is that a viable business model for community energy projects does not exist yet. From the perspective of energy operators, there are external pressures and economic conditions that prevent those business models based around communities from succeeding. In Malawi, some of the most salient challenges include wrestling with systems of governance and regulation and facilitating community participation and leadership.

Decades of experience of community energy projects in Malawi can be capitalized to facilitate an energy future in which community energy becomes ubiquitous. What can be learnt from existing experiences of community energy in Malawi? This chapter presents a comparative analysis of existing community energy projects, focusing on the conditions and challenges of implementation.

Primary data was collected from field visits to five community energy sites in the districts Rumphi, Mchinji, Dedza, Mulanje and Nsanje. A literature review of published and grey literature complemented the dataset. In every case, community energy has proven to play an important role in facilitating access to electricity, and the projects are presented as fundamental to facilitate community and household-level development. However, the projects face multiple challenges both in terms of adapting to governmental regulations and in managing the process of community involvement. The analysis provides policy recommendations, especially the need to support communities directly to facilitate the adaptation of energy systems to different locations, including explicitly both domestic and productive uses of energy and providing access to resources for systems development, such as subsidies.

9.2 Community Energy and Energy Access Challenges in Malawi

Social development depends on having energy access, including both access to electricity and fuels (Casati et al., 2023). In Malawi, the access rate to electricity (14.2% in 2021, and one of the lowest in the world) hinders people's well-being and livelihood opportunities (ESMAP, 2021). The rates of access become dismal when considering only the rural population (5.6% in 2021) (SEA4ALL, 2021). The challenge of accessing electricity is compounded by the challenge of accessing clean fuels because only 2% of the population has reliable access to clean fuels (ESMAP, 2021). This means that almost all of the Malawian population depends on firewood and charcoal for heating and cooking (NCS, 2017), with a direct impact on the health and well-being of communities in both rural and urban areas and high rates of deforestation that threaten the ecosystems of the country. In Malawi, the trade-offs between energy access and ecosystem health raise complex challenges for the delivery of energy justice that any thinking of transition needs to consider (Grant et al., 2021).

The energy sector is characterized by low electricity generation capacity, estimated at about 522.55 MW from all installed units, with the national generation utility, EGENCO, supplying 441.55 MW and two solar-independent power producers contributing an additional 81 MW (JCM, 2022). Approximately 88% of the total electricity supplied by EGENCO is from hydropower stations cascaded along the Shire River, a situation that is considered very risky, for example, in the case of drought or flooding (EGENCO, 2023). EGENCO has faced several setbacks, including the damage caused by Tropical Cyclone Ana that paralyzed its 129 MW hydropower plant at Kapichira Power Station in 2022. EGENCO's ambitious expansion plans were halted once again by cyclone Freddy in 2023, but it has recently been energized by solar investments (in November 2023, EGENCO began a new 10-megawatt Solar Power Project in the township Salima), in line with its 2023–2038 Strategic Plan to increase power generation capacity and diversifying power sources.

Low generation capacity is compounded by limited transmission and distribution facilities and, subsequently, frequent power outages.

9.2.1 Histories of Energy and Community in Malawi

Before independence, the energy developmental initiatives in Malawi were mostly introduced by white missionaries who arrived in Malawi from different countries, especially Britain. Missionaries also led the development of the first energy schemes and decentralized systems in the early twentieth century). For example, a missionary-led project in the Matandani Adventist Mission in Neno (Mission to Matandani, 2021) a micro-hydropower system with 28 kW installed capacity, supplied power to the entire mission, including staff houses, school premises and offices, for more than three decades (MCEM, 2015). This, like other similar systems, fell into disrepair after the departure of the missionaries from the district due to a lack of proper management and maintenance.

However, the establishment of public utilities during the colonial and post-independence periods moved attention away from off-grid systems, and attention to community energy waned. In Malawi, the need for electricity to access electric appliances such as phones, radios and TVs pushed and motivated communities to find their own ways of generating and accessing electricity in diverse ways from solar home systems to micro-hydropower plants. This interest has offered opportunities for investment to international donors and civil society organizations that have seen community energy as opening a new realm for development interventions.

In many countries, such as for example, the Netherlands or Germany, community projects have garnered attention because of their potential to address growing climate change concerns while reducing household energy expenses (Dóci & Vasileiadou, 2015). In Malawi, this is taken to extremes: Malawian communities have been pushed to be energy-creative, resilient and independent by finding ways to generate their own electricity for home consumption and productive use. This is the case of the Kavuzi Pico hydro schemes, where several households have managed to procure and instal pico hydro systems just to meet the most basic household energy needs (cooking and heating are often excluded). Such pico systems are constructed with locally available resources by local artisans from their area. Some organizations have also promoted energy access through community-led energy kiosks (Galichon & Payen, 2017). These provide phone and solar battery charging services and lend solar lanterns at a small fee. For example, the Renew'N'Able Malawi (RENAMA) Rural Off-Grid Energy Kiosks project was launched in 2012 in two districts in the southern region of Malawi, Phalombe and Thyolo.[1] These projects have mainly targeted low-income households, especially in areas where the grid has not arrived, but they often have secondary impacts in neighbouring areas in terms of facilitating the spread of technology alongside new imaginations of energy futures.

Pilot community energy projects, however, took place in an unregulated space. In 2004, the Malawian government implemented six pilot off-grid solar hybrid mini-grids in six districts across Malawi co-funded by local governments. These six mini-grids, with an installed capacity of 20 kW to 25 kW, were designed to supply electricity to households within a one km radius. However, many of these projects were abandoned due to poor management and lack of maintenance and funding. The top-down approach to these projects meant that true community ownership could not be achieved, even when community-based committees were formed to handle the day-to-day administration.

9.2.2 *Current Regulatory Landscape for Community Energy*

Today, most actors in the energy sector recognize that community energy plays an important role in diversification, complementing the significantly larger investment capacity of EGENCO and independent private producers. A conducive regulatory framework has facilitated investments in community energy. Community energy systems have been a common fixture of the energy landscape since 2006, ranging from micro-grids to mini-grids and almost exclusively powered by renewables: solar PV, wind power, hydropower and hybrid systems. The 2018 National Energy Policy aims to provide a guiding framework for increased access to affordable, reliable, sustainable, efficient and modern energy for all sectors and every person in the country (GoM, 2018). It encourages community-level electricity generation and commits government funds to develop rural off-grid generation projects (Eales & Unyolo, 2018).

The 2004 Malawi Electricity Act defines a mini-grid as a system that generates electricity and supplies electricity to a local community with a maximum capacity of 5 MW. Mini-grids are either connected to the main

[1] http://www.renewnablemalawi.org/index.php/projects_field-projects_rural-energy-kiosk-2/.

grid or off-grid, and they are confined to a delimitated area to which they provide electricity. This definition is rooted in the concept of communities' self-production, which is thought to reduce external dependency to facilitate both energy access and security since household energy security depends on having an uninterrupted availability of energy or electricity at an affordable price (Chirp & Jewell, 2014). The 2017 Renewable Energy Strategy puts mini-grids at the centre of the country's vision for the energy sector. It proposes the establishment of Community Energy Service Companies and collaborations with non-governmental organizations such as Malawi Community Energy. The 2017 National Charcoal Strategy promotes the use of alternative fuels and puts Community Development at the heart of the development of alternatives that can help halt deforestation. These regulatory instruments have raised the profile of mini-grids and community energy, have helped recognize them as central to the energy mix and have supported the redirection of investment to these projects, although the development of projects has been fragmented. There is also room for reconceptualizing community energy from a broader perspective, as energy communities; a perspective that emphasizes communities' claims to gain energy sovereignty beyond the capacity of generating electricity (for example, through community-based management of fuels and energy needs).

In addition, some regulatory instruments have been directed towards facilitating the implementation process of mini-grids. The Malawi Energy Regulatory Authority works hand in hand with the Malawi Bureau of Standards to ensure standards compliance in the provision of energy services in the country and, thus, has a strong influence on the implementation of community energy. In July 2020, MERA adopted a Regulatory Framework for Mini-Grids, which recognizes the wide diversity of governance arrangements that enable mini-grids proposing five 'acceptable ownership arrangements': community-based either as a Trust or by cooperative associations, public, private, private–public partnership and hybrid. In addition, the Framework gives communities a mandate to participate in the development of electricity tariffs, for example, ensuring these tariffs are affordable for them.

9.2.3 *The Rural Electrification Program*

The government of Malawi government developed the Malawi Rural Electrification Program (MAREP), following the adoption of the Rural

Electrification Act by the Malawi Parliament in 2004.[2] MAREP aims to provide electricity to rural commercial centres not connected to the electricity grid through various energy options such as solar, hydro and wind, whether they are off-grid or on-grid. The Electricity Supply Corporation of Malawi (ESCOM), a state-owned power transmission and distribution company established in the year 1984, has the mandate to roll out the MAREP program by extending the national grid network to rural communities across the country. The Ministry of Energy, Natural Resources and Mining, EGENCO and Independent Power Producers are supposed to support ESCOM in this electrification work (Fig. 9.1).

MAREP is funded by an energy levy of around 4.5% of energy sales made available to the Rural Electrification Fund (DoE, 2009). The program is being carried out in phases, which depend on the availability of funds. As of November 2023, eight phases of the MAREP program have been completed, and the ninth phase is underway. In this period, ESCOM has electrified more than 108 rural commercial centres (ESCOM, 2023). Despite these efforts and the modest increases in electrification rates, Malawi still faces an enormous energy access challenge.

The government acknowledges that improving access to electricity in rural communities through grid extension may be too slow to deliver universal energy access beyond 2030, especially due to a lack of funding.

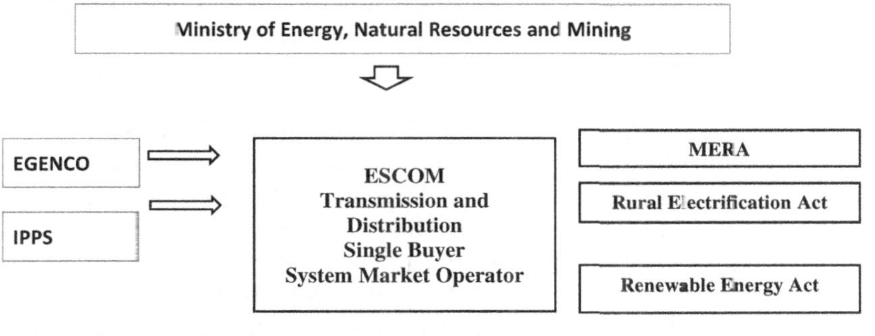

Fig. 9.1 Institutions and regulations that support the Malawi Rural Electrification Program

[2] Rural Electrification Act (2004) Retrieved January 14, 2024, from https://faolex.fao.org/docs/pdf/mlw118920.pdf.

Decentralized energy systems have thus a key role to play in the electrification of rural communities (GoM, 2018). Simpler technologies and nimble systems have facilitated the development of off-grid systems that involve communities in the planning and implementation of energy systems and facilitate community ownership. In Malawi, this strategy may be vital for hard-to-reach areas, especially rural communities that are far from the power grid (Alexandros, 2021).

These efforts, while positive, are not sufficient. The funding gap in energy infrastructure development in Malawi is estimated at US $2.5 billion to increase electricity generation to 1,200 MW (doubling the current 441.95 MW) and connect 1.2 million homes (MoNRENM, 2019). By this calculation, a thriving off-grid sector would require at least US$ 130 of investment funds before 2030. Investment alone will also require the active participation of all stakeholders, especially private investors, communities and civil society, to deliver universal energy access.

9.3 Community Energy in Practice: Five Projects in Malawi

By working together on off-grid initiatives in Malawi, communities, in partnership with NGOs and private actors, are advancing access to energy in rural communities. Governance arrangements are highly variable: some of the off-grid initiatives are fully owned and managed by the communities through electricity village committees with the help of the chiefs. The most frequent challenges with these off-grid systems, which contribute to unsustainability, are thought to be a lack of maintenance and unreasonable, cost-effective tariffs (Eales & Unyolo, 2018).

This section covers the results from the assessment of various community energy systems (CES) initiatives implemented in Malawi by different stakeholders in the energy sector. Both primary and secondary data regarding these case studies were gathered through direct interviews and desk research, respectively. A total of 315 households connected to the mini-grids were surveyed to gain insights into their energy usage patterns, as well as the benefits and challenges they encounter in accessing electricity from CES mini-grids. Additionally, interviews were conducted with at least 5 CES operators or managers at the visited CES sites to comprehend the diverse best practices employed by CES projects throughout Malawi. The selection of these case study sites was based on the accessibility of the site, system size, and the number of beneficiaries involved in each project. Analysis is still ongoing, but this chapter offers a first comparative overview of the implementation of the projects in context, attending to the focus on productive use, the involvement of the community and the main challenges identified in the interviews.

9.3.1 CARD Mini-grids and Kiosks

Churches Action in Relief and Development (CARD) is a humanitarian, church-based NGO running solar off-grid programs in four locations: Nyamvuwu and Chimombo in Nsanje, and Oleole and Mwalija in Chikwawa. The project targets 20,000 beneficiaries, including women and children in four of its mini-grids and kiosks sites. These solar off-grid systems were designed to reduce poverty and increase resilience in the target areas by providing electricity for productive uses. All the systems were commissioned in 2019 and are still in operation today, except for the Oleole mini-grid, which was damaged by floods in 2021 (Fig. 9.2).

Fig. 9.2 Mwalija solar mini-grid

The project was funded by the European Union as part of the Sustainable Energy for Rural Communities project (SE4RC), with an initial investment of ca. $840,000. SE4RC was a collaboration between Practical Action, HIVOS and Environment Africa. SE4RC also set up seven energy kiosks to promote and improve access to clean and affordable energy for low-income households in grid-hard-to-reach areas. SE4RC strategy was to train community members to constitute Community Energy Service Companies (CEOSCs) that can become legal entities to access funds and provide sustainable services to the local communities. These CEOSCs own, operate and maintain the mini-grids. The CEOSC performs several functions, including minor maintenance on the Mini-grid system under the supervision of the Malawi Energy Regulatory Authorities.

The connection fee for these mini-grids requires the user customer to commit an amount of approximately $25.00, plus meeting the cable wiring costs, which include procurement of electrical wires. These energy systems adopted the pay-as-you-go business model to collect revenue from users connected to the system. However, their vending systems are no longer operative due to a system server problem, which required further investments of at least $1000 to get the vending system up and running. After the vending system stopped working, CEOSCs established a flat rate of about $3 monthly per household. Collected revenue is used to run the mini-grid and procure equipment, such as cables and new household connections with some additional funds used for community initiatives.

9.3.2 Sitolo Solar Mini-grid

The Sitolo solar mini-grid in the village of Sitolo, Mchinji district, is one of the first of its kind in Malawi. The installation was implemented by Community Energy Malawi (CEM) in 2019 after receiving a $435,000 investment from UNDP. This mini-grid generates 80 kWp of electricity and operates 24 hours a day without any backup diesel generator. It provides electricity to 1500 households.

CEM owns, maintains and operates the mini-grid. Each household pays a $30 connection fee. A big inspiration for the project was the Mulanje Electricity Generation Agency (MEGA) mini-grid (see below and Chapter 10) and, like MEGA, it operates as a social enterprise focusing on making electricity available and affordable to the community/target market by minimizing prices. To ensure the mini-grid is sustainable,

CEM employed an experienced technician to oversee system maintenance, troubleshooting and on-site technical supervision.

9.3.3 Bondo Mini-grid

Bondo Mini-grid includes several small hydroelectric power stations built along the Lichenya River. These power plants are located in Bondo Village, Mulanje District, in the southern part of Malawi, and have been operational since 2012. The project was initiated by the Mulanje Conservation Trust to address the problem of deforestation around Mulanje Mountain. The funding for the mini-grid infrastructure was provided by the European Union, the Scottish government and the government of Malawi.

Subsequently, Mulanje Electricity Generation Agency (MEGA) was established as a business entity to manage the mini-grid as a social enterprise. MEGA became the first independent power producer in Malawi to obtain a power generation and distribution licence from the Malawi Energy Regulatory Authority. During the first phase of the project, the local communities were involved in the installation process through job opportunities. The community also procured land for system installation but the land for the transmission lines also required consultation with traditional leaders. During the trial phase, twelve houses were connected to the grid. This sparked a demand from the entire community and a further integration of MEGA within the community.

In 2023, the mini-grid had connected over 2000 houses across three villages. Customers are connected to the grid through prepaid metres, with three tariff categories: commercial at $0. 9/kWh, residential at $0. 7/kWh, and social at $0. 4/kWh. Plans to expand the mini-grid to other villages depend on the availability of funds but advance apace.

9.3.4 Mthembanji Micro-grid

Mthembanji micro-grid was implemented as part of the Rural Energy Access and Social Enterprise (EASE) project in Dedza District, Central Malawi. EASE was a 1.3 million euro-funded Scottish Government project to promote decentralized energy access in Malawi via a partnership between the University of Strathclyde, Community Energy Malawi, United Purpose and the Centre for Water Sanitation, Health and Appropriate Technology Development (WASHTED) (Fig. 9.3).

Fig. 9.3 Mthembanji solar PV mini-grid

The project also operates under a social enterprise model. The microgrid started operating in 2020. In 2022, at least 53 households, five tradespeople and two institutions were registered with the customer registration. There are different tariff categories to suit every level and type of energy demand in the Mthembanji community. One of the categories is the Banya package, which requires a customer to pay $4 per month. The other one, the Ufulu Bundle PAYG tariff, depends on the time, the type of usage and the usage level, while the Mudzi package is aimed at institutions and paid monthly.

The project employed two local agents trained in essential system maintenance, sales and customer relations. However, the EASE project budget covers part of the operation and maintenance costs, potentially raising the alarm about the system's sustainability beyond the funding period.

9.3.5 *Chipopoma Mini Hydropower*

Chipopoma is a community-based micro-hydropower mini-grid in the Rumphi district that supplies electricity to more than 120 houses with

1500 direct individual beneficiaries. The system also provides power to a nearby primary school, health centre and orphan care. The micro hydro power plant generates approximately 50 kW transmitted through an 11 kV medium voltage grid that is 3.5 km long. Distribution lines extend an additional 9.5 km at 230 V. The idea to harness hydroelectric power from the Mantchewe waterfalls began in 2015 when a village resident mobilized the community and sourced funding from a private business, the Mushroom Lodge, to procure PVC pipes for penstocks. The turbine and generator were constructed and assembled locally using locally available resources. Following the success of this initiative, the community energy project secured additional funding from UNEP to instal a standard transformer and transmission and distribution lines to connect more households.

The plant provides energy for ca. 100 households. The hydropower plant is operated by a group of managers who regularly consult the community through the institution of the Board, with which all decisions are consulted. The majority of the Board members are women. The project plans to expand the generation capacity to 120 kW to accommodate more households and facilitate revenue generation.

9.4 Productive Uses of Electricity

The first question asked comparatively was the importance of productive uses of energy in each site. Productive Uses of Electricity (PUE) add value to electricity through income-generating activities. Harnessing PUEs is a strategy to add demand to energy systems and make them financially viable while also expanding livelihood opportunities within a community (Vivien et al., 2019). Providing rural communities access to sustainable electricity significantly impacts community and household economies (Peters & Sievert, 2016). The experiences of community energy in Malawi show that they foster economic activities, ranging from small to large business enterprises.

For example, a key sector of interest is subsistence agriculture, still the rural population's primary livelihood source. However, low access to electricity reduces the opportunity to increase the value of agricultural produce.

In all projects visited, one common PUE was refrigeration to facilitate the sale of cold drinks, frozen food and fresh fish. The availability of electricity enabled shops to extend their opening hours and increase sales

(Fig. 9.4). Hairdressing businesses, including barber shops and salons also used electricity. Sometimes electricity is used for domestic activities that also facilitate businesses such as phone and car battery charging. Other times electricity enables industrial manufacturing, for example, the fabrication and manufacture of building materials like window and door frames.

The type and size of the system and its location conditions the PUEs possible in each site. For example, maize mills were not commonly found in CES powered by solar PV because they tend to have limited generation capacity. The Sitolo Mini-grid had a maize mill and an oil refinery plant that operated on a prescribed schedule to regulate the electricity demand. CES sites in Chikwawa and Nsanje, Nyamvuwu, and Malija, had successful irrigation schemes serving at least 250 beneficiaries. These schemes have contributed to improved intercropping practices, thereby increasing food security and income levels. Our comparative assessment suggests that electricity has brought a host of benefits to these communities apart from household electrification, mainly through PUEs, but that PUEs are also fundamental to the constitution of community energy and, therefore, need to be put at the centre of its development.

Fig. 9.4 Using refrigeration services in Chipopoma

9.5 Community Participation

The success of community energy depends on the active participation of the community. The most difficult aspects to negotiate with the community are the financial sustainability of the project and its governance. With regard to the financial sustainability of community energy, all projects require communities to show a willingness to pay for energy services, but this may not always be readily accepted. A participatory process can help negotiate fair tariffs between communities and system operators (which may overlap but are rarely the same). The same process may be deployed to engage the community in the rehabilitation of the systems and day-to-day management.

The governance needs of community energy pose their own challenges. In Malawi, community energy is often owned by the communities to various degrees. However, what 'ownership' means in practice varies hugely and depends on various factors, such as the business model, how it was introduced to the community, and the decisions made by the community regarding its operation.

The five projects visited show that full ownership by the communities entails strategic leadership to day-to-day management. For example, the four CARD solar mini-grids are fully owned and operated by the communities, whose members have taken responsibility for making new connections, expanding the grid and recruiting additional expertise. The Bondo community had concerns about managing their community energy system because they had unrealistic expectations that electricity should be provided free of charge. These unrealistic expectations were generated during the introduction of the project. In contrast, the Mthembanji mini-grid, which is owned and operated by United Purpose, does not plan to hand it over to the community due to the complexity of community ownership in relation to the business model used by the community energy system.

Regardless of the level of ownership, two features are common to all the projects to guarantee purchase by the community. One of those features was the presence of a village electricity committee responsible for each energy system. These committees are chaired by traditional leaders or chiefs and serve to control and address any community grievances regarding the services provided by the community energy systems. Chiefs are often involved in decision-making processes related to the systems because they are thought to represent the voice of the community (an assumption that merits interrogation from an intersectional perspective).

The other feature that makes participation in community projects successful is the involvement of the communities in providing land for the installation of the community energy systems. For example, in Mthembanji village, the community granted part of the land while the other remaining was leased through the Ministry of Lands, with consultations and approvals from traditional local leaders. In the other four sites, land was given by the community in consultation with the local leaders.

9.6 Challenges to the Development and Maintenance of Community Energy

The most common challenge for all these projects is their sustainability over time. Most community energy projects are actively supported and promoted in the initial years of funding. However, the continued provision of service after funding stops is unwarranted. Managers in all visited sites mentioned funding as the major challenge faced which hindered the expansion and maintenance of their projects.

Tariff revenue is often insufficient to meet community energy's daily running costs. For example, the CARD Mini-grids project faced financial and management challenges after transferring ownership to the community to handle the project's day-to-day management. The local electricity committee responsible for executing the project was trained on the best practices for its implementation. However, shortcomings in management were discovered along the way, specifically in the misuse of funds. The committee has been unable to fund further household connections because they lack the necessary funds to acquire materials, and they could not face any technical difficulties, breakdown of services, or system failures with the existing resources. In Chipopoma, the management team has gone on weeks or even months without services as they tried to put together the funds to repair the network.

Many managers emphasized that communities felt deeply a lack of technical expertise. While local technicians from the community may have expertise, complex equipment failures require specialized knowledge to repair. In the case of the Oleole mini-grid in the Chikwawa district, floods damaged the inverter and charge controllers. The trained CESOES lacked the expertise to fix this equipment and hired an external expert to repair the system, which further challenged their available funds. One inverter for the Mthembanji mini-grid had been nonfunctional for months because the local community members lacked the expertise to repair it.

Many of these challenges stem from the fact that despite the national-level efforts, these community energy projects depend on inadequate policies and regulatory frameworks. According to operators, the policy environment does not support community energy projects to thrive. Collaboration with ESCOM on rural electrification or grid integration is rare. In the case of the Nyamvuwu Mini-grid, the MAREP program through ESCOM had encroached on the area, and this has made the customers shift their connections from the mini-grid to the grid. While this may be good news for the customers (who may find the grid more reliable or more conventional) it represents a disaster for a stretched community energy project that depends on meagre tariffs. Despite the celebration of community energy, national policies do not reveal key details, such as for example, how can independent community energy projects access governmental funds. whether CES has the right to access funds from the rural electrification fund. None of the projects visited had received governmental funds.

Establishing effective revenue collection methods in mini-grid systems can present challenges if proper processes are not defined. For instance, in CARD mini-grids, where the vending system stopped working due to server challenges and in Chipopoma, which lacks an automated vending solution, collecting monthly revenue becomes difficult as some customers may not pay their electricity bills. In Bondo, customers must walk to the operator's office to purchase electricity tokens, with a waiting time of one to two days for tokens to be available. A similar situation exists in Sitolo and Mthembanji, where customers also walk to offices to buy tokens, requiring potentially long distances to be travelled. Across all sites, customers cannot purchase electricity through mobile money or banking services.

These insights have been summarized in Tables 9.1 and 9.2. While Table 9.1 explains the factors that enable community energy in Malawi, Table 9.2 turns attention to the challenges faced in these projects. Table 9.1 demonstrates that these projects depend on a combination of external support (i.e., funding) and the integration of the project into the local community, particularly through productive uses of energy. However, most of the external support has come from international development organizations and NGOs. The intervention of the government along the lines of existing policies could mobilize public funding in the form of grants to fund the next generation of community energy projects but also to support existing projects facing hardship on their road towards sustainability.

Table 9.2 in contrast provides an overview of the diverse challenges that these projects face, in the technical arena, the availability of materials through the supply chain and the numerous uncertainties- some related to natural disruptions beyond the control of the operators. Community participation, ownership and a number governance structure may all help to support these valuable projects over time.

9.7 Conclusion

Several important lessons must be learned from community energy systems in Malawi. The government policies promote off-grid renewable energy but lack clarity and specific support for community projects. This chapter suggests a need for a comprehensive review of energy policies to accommodate community energy systems to benefit from the rural electrification fund and other government subsidies, such as user-targeted subsidies.

Based on the empirical data, the study found that technical training was provided to some community energy system operators by the government through the Malawi Energy Regulatory Authority, Department of Energy Affairs, Mzuzu University and other stakeholders. This training positively enhanced the maintenance challenges faced by community energy. However, financial support is needed for many projects that have fewer customers and are in their early implementation stage. Promoting community stewardship by involving traditional leaders has been found to be an effective approach, especially in rural settings, but their role in shaping the process of social inclusion needs to be interrogated. Transferring full ownership of community energy to the communities requires careful evaluation and planning to achieve and particularly, avoiding the generation of unrealistic expectations.

The most important factor overlooked in the operation of community energy is the resilience of different projects and how they respond to uncertainties and shocks (see Chapter 2). Resilience can be incorporated into project design with the objective of preventing system failures. Better environmental assessments can also help community energy projects to endure external disruptions.

Table 9.1 Key aspects in the delivery of community energy in Malawi

Community Energy System	*Combination of Official Development Assistance & funding sources*	*Transfer of ownership to the community*	*Local governance/ Institution Support*	*Employment opportunities for the locals*	*Market Expansion*
Sitolo Solar Mini-grid	Funded by UNDP, Land leased from the community	Community engagement and partial ownership rights	Training support from local organisation academic institution on PUE and Maintenance Policy support from MERA and DoE	Provision of permanent and temporary job opportunities; construction, System Maintenance, Marketing	Various Productive use of electricity activities by locals
Mnthembanji Solar mini-grid	Funded by United Purpose, CES Land leased from the community	The CES is being run by United Purpose, no plans of transferring the ownership to the community	Training support from local organisation, Policy Support from MERA and DoE Knowledge sharing	Provision of permanent and temporary job opportunities; construction, System Maintenance, Marketing	Various Productive use of electricity activities by locals
Bondo Mini-grid	Funded by Donors, CES Land donated by the community	MEGA runs the CES in collaboration with local committee members which also includes chiefs	Training support from local organisation, academic institution on PUE and Maintenance, Policy support from MERA and DoE, Knowledge sharing on the best practices	Provision of permanent and temporary job opportunities; construction, System Maintenance, Marketing	Various Productive use of electricity activities by locals

(continued)

Table 9.1 (continued)

Community Energy System	*Combination of Official Development Assistance & funding sources*	*Transfer of ownership to the community*	*Local governance/ Institution Support*	*Employment opportunities for the locals*	*Market Expansion*
Card Mini-grids	The mini-grids were fully funded by SE4RC project, CES Land donated by the community	The 4 mini-grids are now fully owned and managed by the communities	Training support from local organisation, academic institution on PUE and Maintenance, management, entrepreneurship, Policy Support from MERA and DoE	Provision of permanent and temporary job opportunities; construction, System Maintenance, Marketing	Various Productive use of electricity activities by locals
Chipopoma Hydro Mini-grid	Community initiative with joint funding from Mushroom lodge, UNDP fund for distribution lines	CES fully owned and operated by the community	Training support from CESET project and Mzuzu University	Provision of permanent and temporary employment opportunities; construction, System Maintenance, Marketing	Various Productive use of electricity activities by locals

Table 9.2 Challenges to the implementation of community energy in Malawi

CES Type	*Availability of equipment/ Supply Chain*	*Technical & Operation*	*Natural Disruptions*	*Policy & Framework Barrier*	*Funding*
Sitolo Solar PV Mini-grid	The majority of the equipment is imported	the CES fails to maintain by replacing larger worn-out system components due to funding	No Major natural event that ever disrupted the operations Heavy rainfalls and cloudy weather affect the generation	No specific policies on grid integration	Lack of additional external funding to expand its system
Mnthembanji Solar PV Micro Grid	Most of the equipment are imported	Does not have the capacity to maintain the system without funding support	No Major natural event that ever disrupted the operations	No policies that allow CES to benefit from RE fund	No extra funding required as of now
Bondo Mini-hydro Mini-grid	Most of the equipment are found locally apart from the turbine and generator	revenue collected not enough to support large scale system maintenance	Bondo Power house was destroyed by floods and killed 2 watchmen	No policies that allow CES to benefit from RE fund	Limited access to financing to expand its system and connect more customers

(continued)

Table 9.2 (continued)

CES Type	*Availability of equipment/ Supply Chain*	*Technical & Operation*	*Natural Disruptions*	*Policy & Framework Barrier*	*Funding*
CARD Solar PV Mini-grids	Most for the equipment such as batteries, inverters are imported from China	revenue collected not enough to support large scale system maintenance due to funding, lack of Managerial skills	Mwalija Solar Mini-grid was affected by floods which damaged the equipment except the solar panels	No specific policies on National Grid encroachment, No formal collaborations with Mini-grid developers	Require additional funding to expand its system and connect more customers
Chipopoma hydro Mini-grid	All the equipment used for construction were sourced and fabricated locally	the CES fails to maintain by replacing larger worn-out system components due to funding	*No Natural Disruption*	No policies that allow CES to benefit from RE fund	Limited access to financing to expand its system and connect more customers

One advantage of focusing on PUEs is the possibility of maximizing electricity usage during off-peak hours, a strategy that can help improve revenues and project survival rates. Other strategies include encouraging and promoting productive use of energy through cooperatives, such as irrigation farming or small-scale agro-processing. Community energy systems that have been well established have the potential to contribute meaningfully to rural development in Malawi. Sustainable community energy can help drive economic and social development in underserved rural areas.

References

Alessi, P. L. (2022). *Access to modern energy and the impact on health*. Retrieved January 5, 2024, from https://energypedia.info/wiki/Access_to_modern_energy_and_the_impact_on_health

Alexandros. (2021). *How do we power rural communities? By providing off-grid solutions*. Retrieved from https://odi.org/en/insights/how-solar-mini-grids-can-bring-cheap-green-electricity-to-rural-africa

Barnier, V., Boursier, S., Holzigel, J., Kamali, J., Majumder, D., Peterschmidt, N., Browne, C., Nderitu, D., Bharadwaj, A., & Weston, P. (2019). *Demand side management for mini-grids*. Retrieved January 5, 2024, from https://energy4impact.org/demand-side-management-mini-grids-0

Casati, P., Moner-Girona, M., Khaleel, S. I., Szabo, S., & Nhamo, G. (2023). Clean energy access as an enabler for social development: A multidimensional analysis for Sub-Saharan Africa. *Energy for Sustainable Development, 72*, 114–126.

Cecelski, E., & Glatt, S. (1982). *The role of rural electrification in development*. Resources for the Future.

Cherp, A., & Jewell, J. (2014). The concept of energy security: Beyond the four As. *Energy Policy, 75*, 415–421.

Chisale, S. W., & Lee, H. S. (2023). Evaluation of barriers and solutions to renewable energy acceleration in Malawi, Africa, using AHP and fuzzy TOPSIS approach. *Energy for Sustainable Development, 76*, 101272.

Dóci, G., & Vasileiadou, E. (2015). 'Let's do it ourselves'. Individual motivations for investing in renewables at community level. *Renewable and Sustainable Energy Reviews, 49*, 41–50.

DoE (2009) *The Malawi rural electrification promotion project*. Japan International Cooperation Agency.

Eales, A., & Unyolo, B. (2018). *Renewable energy mini-grids in Malawi: Status, barriers and opportunities*. Retrieved January 5, 2024, from https://strathprints.strath.ac.uk/64868/

EEP (2018). *Opportunities and challenges in the minigrid sector in Africa*. Retrieved January 5, 2024, from https://eepafrica.org/wp-content/uploads/2019/11/EEP_MiniGrids_Study_DigitalVersion.pdf

EGENCO (2023). *Electricity Generation Company (Malawi)*. https://www.egenco.mw

ESCOM (2023). *MAREP Department*. Retrieved January 5, 2024, from https://www.escom.mw/marep.php

ESMAP (2021). *Tracking SDG7*. Retrieved January 5, 2024, from https://trackingsdg7.esmap.org/country/malawi

Galichon, I., & Payen, L. (2017). *Energy access in rural Togo: the relevance of the energy kiosk*. Retrieved January 5, 2024, from https://inis.iaea.org/search/searchsinglerecord.aspx?recordsFor=SingleRecord&RN=48094814 2017

GoM (2018). *National energy policy*. Retrieved January 14, 2024, from https://npc.mw/wp-content/uploads/2020/07/National-Energy-Policy-2018.pdf

Grant, R., McCauley, D., Von, M. M., Grattage, R., & Mwathunga, E. (2021). An Ecohealth approach to energy justice: Evidence from Malawi's energy transition from biomass to electrification. *Energy Research & Social Science*, 101875.

Grivas, A. (2021). *How do we power rural communities? By providing off-grid solutions*. Retrieved January 5, 2024, from https://www.stantec.com/en/ideas/topic/stantec-era/how-do-we-power-rural-communities-by-providing-off-grid-solutions

IRENA. (2022). *Renewable energy statistics*. Retrieved January 5, 2024, from https://www.irena.org/Publications/2022/Jul/Renewable-Energy-Statistics-2022

JCM. (2022). *Project overview*. Retrieved January 5, 2024, from https://www.jcmpower.ca/salima

MCEM. (2015). *Increasing access to clean and affordable decentralised energy services in selected vulnerable areas of Malawi*. Retrieved January 19, 2024, from https://www.undp.org/malawi/projects/increasing-access-clean-and-affordable-decentralised-energy-services-selected-vulnerable-areas-malawi

MERA. (2020). *Regulatory Framework for Minigrids*. Retrieved January 5, 2024, from https://mera.mw/downloads/regulatory-framework-for-mini-grid-july-2020/

MERA. (2023). *Company overview*. Retrieved January 5, 2024, from https://mera.mw/overview/

Mission To Matandani. (2021). Retrieved January 5, 2024, from https://echo.sid.adventist.org/mission-to-matandani/

MoNRENM. (2019). *Malawi sustainable energy investment study*. Retrieved January 14, 2024, from https://www.un.org.ohrlls/files/malawi-sustainable-energy-investment-study.pdf

NCS. (2017). *National charcoal strategy 2017–2027*. Retrieved January 5, 2024, from https://cepa.rmportal.net/Library/government-publications/national-charcoal-strategy-2017-2027/view

NSO. (2019). *2018 Malawi population and housing census*. Retrieved January 5, 2024, from https://malawi.unfpa.org/sites/default/files/resource-pdf/2018%20Malawi%20Population%20and%20Housing%20Census%20Main%20Report%20%281%29.pdf

Peters, J., & Sievert, M. (2016). Impacts of rural electrification revisited–the African context. *Journal of Development Effectiveness, 8*(3), 327–345.

Puranasamriddhi, A., & Parikh, P. (2021). *Off-grid energy and economic prosperity*. Retrieved January 14, 2024, from https://www.ucl.ac.uk/bartlett/construction/sites/bartlett_construction/files/00impact_of_off-grid_in_ssa_report_final.pdf

SEA4ALL Malawi. (2021). Retrieved January 5, 2024, from https://www.se4all-africa.org/seforall-in-africa/country-data/malawi/

UNDP. (2021). *An overview of community energy systems in Malawi*. Retrieved January 19, 2024, from https://cesetproject.com/sites/default/files/An%20Overview%20of%20CES%20in%20Malawi_Nov21.pdf

Vivien, B., Sylvain, B., Joanis, H., John, K., Majumder, D., Browne, C., & Bharadwaj, A. (2019). *Demand side management for mini-grids*. Retrieved from https://energy4impact.org/sites/default/files/afdb_gmg_dsm_report_final_draft_cb.pdf

CHAPTER 10

Learning from the Ground Up: Community Energy Stories from Malawi

Arnold Kadziponye and John Sailence

Abstract There is a need to engage directly with community energy managers and operators to co-produce knowledge directly relevant to the challenges they face. In this challenge, two operators of community energy in Malawi relate their experiences on the ground, making community energy projects possible. This chapter presents two accounts of the making of community energy in Malawi, as told by the protagonists of those stories.

The first experience, written by Arnold Kadziponye, reports the story of the Mulanje Electricity Generation Agency Ltd (MEGA), in the Mulanje Massif in the south of Malawi (65 km from Blantyre). Arnold's story shows how concerns with conservation foster the development of an electrification project. Now covering over 2000 households and expanding rapidly, the biggest lessons of Arnold's stories relate to the

A. Kadziponye (✉)
Mulanje Energy Trust, Mulanje, Malawi
e-mail: arnold@mountmulanje.org.mw

J. Sailence
Chipopoma Power, Livingstonia, Malawi

V. Castán Broto (ed.), *Community Energy and Sustainable Energy Transitions*, https://doi.org/10.1007/978-3-031-57938-7_10

engagement with the community and the need to moderate unrealistic expectations.

The second story presents the Chipopoma Power Project in Manchewe, Livingstonia, in the North of Malawi, and it is reported in the form of an interview with John Sailence, that in 2023 served 120 households. A project assembled with heterogeneous technologies and resources—including learning to build a turbine—demonstrates the potential to harness and develop the knowledge skills within the community.

Keywords Community energy · Social enterprise · Rural electrification · Malawi

10.1 Introduction

This chapter offers two contrasting stories from community operators who explain their experiences with community energy in the first person, paying attention to how local electrification has changed their communities.

The first story is told by Arnold Kadzinponye, who reports on the achievements of Mulanje Electricity Generation Agency Ltd (MEGA). A historical account of the project enables a reflection on how community energy enables place-based transformations. MEGA has grown to serve over 2000 households, but this has entailed overcoming social and economic barriers since its beginnings in 2012. One important challenge relates to the engagement of the community and the management of unrealistic expectations, such as, in this case, the provision of free electricity. MEGA shows that the governance of energy requires institutional development, a process of compromising and enrolling powerful actors within the community. Therefore, the main challenge for MEGA is to ensure that less powerful voices are not ignored and excluded from the benefits of electricity.

The second story is told by John Sailence, who reports through an interview on the Chipopoma Hydropower Project, of which he is the principal architect. The Chipopoma project is a makeshift project that has been painstakingly assembled from multiple resources simultaneously and has faced existential challenges regularly since its inception in 2017.

The case of Chipopoma challenges common ideas about the lack of skills within communities and demonstrates that these can be developed through the construction of a community energy project.

While the projects are different and in different stages of development, they demonstrate the need to read community energy projects as work in progress, always in the making. Their dynamic nature means the project can constantly adapt to new demands and emerging opportunities. Community energy projects may be heterogeneous and sometimes precarious. Still, they help harness local ingenuity and match it with external resources while confronting the radical uncertainty of living under climate change.

10.2 Finding Inspiration in the Story of MEGA

The topography of Mulanje provides good potential for the use of hydropower. The slopes of Mount Mulanje provide high head, while high rainfall rates at elevated altitudes provide significant water flow. The average annual rainfall of Mulanje is 1600 mm, reaching 2849 mm at elevated altitudes (compared to a world average of 860 mm). Surface runoff is the primary water source in the district. The mean annual runoff of Mulanje Rivers is estimated at 51.63 m3/s (including the Rivers Likhubula, Thuchila, Likulezi, Phalombe, Sombani, Nandiwo, Muloza, Lichenya and Ruo). Because of the variability and evaporation losses, only about 66% of the mean annual runoff can be exploited to produce electricity with the current technology. Lujeri Tea Estates were already exploiting some of the hydro potential of the Mulanje Mountain. The estate was generating 840 kW. Much more can be achieved by exploring other rivers, especially in areas far from the national grid (Fig. 10.1).

Mulanje Renewable Energy Agency, Practical Action and Mulanje Mountain Conservation Trust came together to improve access to modern energy in the rural area of Malawi by tapping the potential of the water resources of Mount Mulanje.

> When this initiative was brought into our village, I didn't think I would live to witness the use of electricity in our village, let alone in my house and shop. There was a time when we lost hope that this was impossible; the project was just trying to experiment on us, and it was becoming difficult for the implementers to concede failure, looking at the long time it was taking and the information we had earlier. This brought much desperation.

Fig. 10.1 The rivers of mount mulanje provide ample opportunities for electricity generation

> And now that I am one of the beneficiaries of this, as I do have electricity that I am using to light my house, my electric kettle is used to make hot water within the shortest possible time as opposed to when we had to light a fire using firewood. Imagine I have now procured a refrigerator, which I am stocking cold drinks for sale. (B.B. Godfrey of Chuma ndi Anthu Grocery)

Night falls early on Mount Mulanje. At 17:30 hours, the sun sets, and the local schools in Bondo village (Kabichi Community Day Secondary School and Kabichi Primary School) are in total darkness. After a day's farming at the local smallholder tea field or working at a nearby (15 km away) diesel-powered maize mill, the people of the village light their homes with candles, log fires, kerosene lanterns or battery-powered lamps. While these energy sources provided families with one or two hours of light in which to read and cook, they were becoming increasingly expensive because of high inflation. People in Mount Mulanje felt a pressing need to access electricity.

Lack of access to energy services had other implications for the local communities around Mulanje Mountain. It was rare, for instance, to find qualified teachers and health personnel willing to live in an area without electricity. In Bondo village, schools had to turn down computers from the Ministry of Education and other well-wishers because there was no electricity to run them. Even the local health centre was resource-constrained to the point that women in childbirth were asked to bring their own candles to light their delivery process. Minor ailments such as BP testing were referred to the 25 km away Mulanje District Hospital. The Bondo Health Centre failed to administer quality service as the facility had no storage for medicine because most medicines must be kept in cool places.

The forest reserve was not spared. Almost 100% of the households in the area relied on biomass in the form of firewood for cooking. This firewood was sourced from the Mulanje Mountain Forest Reserve and the degradation exerted to the environment was witnessed each passing day. Efforts by organizations like Mulanje Mountain Conservation Trust to restore and conserve biodiversity couldn't match the levels of degradation caused by the ever-growing population. The land holding size in Mulanje is very low because a big portion of the land is occupied by the tea companies.

10.2.1 The Social Enterprise Model

Mulanje Electricity Generation Agency (MEGA) Ltd is an innovative, inclusive business aiming to transform this situation for households living in mountainous areas with fast-flowing water, where there is potential for micro-hydro power generation. MEGA is a start-up company providing energy to off-grid low-income households. It targets a potential market of 520,000 people in the Mount Mulanje area who will be able to access energy services, of which 9,600 households (42,420 men, women, and children) are expected to be directly connected (Fig. 10.2).

MEGA was established as a social enterprise and is the first operational private energy company in Malawi to be licenced for generation and transmission. MEGA's business model focuses on making energy available and affordable to its target market by promoting price minimization, rather

Fig. 10.2 MEGA is the first social enterprise to be licenced for the generation and transmission of electricity in malawi

than 'traditional' profit maximization, within the parameters of building a financially sustainable business.

History has it that MEGA is the first privately owned micro-hydro energy provider in Malawi. It operates through partnership with MMCT (governance), MuREA (site & business development) and Practical Action (technical expertise). MEGA produces and distributes micro-hydroelectricity to consumers found in off-grid, pro-poor subsistence farming communities at the base of the south-eastern part of Mulanje Mountain. MEGA collects tariffs through prepaid metres to households and credits sold by community vendors (still in progress) and supports the development of business centres and community facilities that may agree on tailored payment plans with MEGA. Consumers are classified according to their energy demands, and these include On-Grid and Off-Grid Rural Households, which fall under the Domestic Tariff; Business entities under the Business Tariff (mostly maize mills but also welders); and lastly, Community Assets (schools, hospitals, churches), under the Social Tariff.

All project preparatory works, including community engagement, feasibility studies, environmental impact assessments, project design and execution, were done by MuREA between 2008 and 2013. MuREA is a non-profit organization established at the end of the Programme for Basic Energy Conservation (PROBEC), a SADC regional programme supported by the German Agency for Technical Co-operation (GIZ). PROBEC had different programmes, including the promotion of clay stoves in rural areas and the training of stove producers, partner organizations and village-based trainers; training metal workers to become institutional rocket stove producers; identifying employers with a large staff base so as to encourage the use of efficient, fixed household rocket stoves (like Esperanza and Changu stoves) and encouraging the development of energy devices that are appropriate for productive use by small and medium enterprises that involve the burning of wood, including the promotion of efficient Tobacco rocket barns.

MuREA took over these programmes from PROBEC with a continued focus on efficient stoves. However, looking at the many perennial rivers flowing from the beautiful Mount Mulanje, MuREA felt it possible to generate hydropower and conducted studies to understand the flow rates and terrain suitable for electricity generation. In 2008, the Lichenya River was identified as a potential site, so the Bondo community was targeted for the project. This was a farfetched dream requiring new technical and social skills to convince the community. The bottom line was that MMCT, being a biodiversity conservation-conscious firm, had an interest in seeing the communities use the generated power for cooking to reduce the degradation of the environment happening on the Mount Mulanje Biosphere Reserve.

People surrounding the Bondo area, drawn from the villages Kalamwa, Nkundi, Nkhulambe, Bondo, Bondo II, Kashoni, Naimbele and Naluwade, were mobilized for the project and contributed to the works in kind through the provision of labour, fetching rocks and sand, and working on the construction to ensure they got electricity. Members of the community became excited upon seeing that power came from an initiative they took part in. The first customer got connected in 2013 when people couldn't still imagine their dream was coming into reality. Seeing the health centre illuminated had an enormous impact on the community. The excitement and jubilation expressed through singing and dancing overnight was an experience that left a mark on each face of the inhabitants of Bondo. This moved many people into investing in having their own houses wired in anticipation that they would soon forget using the hazardous kerosene for some clean, renewable energy.

10.2.2 Building Local Leadership

Community leaders played a very important role in ensuring that community members participated in the project. They developed a roaster showing which village was supposed to work on which date, and they made sure people were working every day, including weekends, to meet the target of delivering the project and having electricity in 5 years. A member of the community GVH Bondo recalls:

> When officers from MUREA came to discuss the potential of generating electricity from the river in our community, we thought it was a lie and that it was in no way possible as we knew MUREA as an organization that promoted efficient cookstoves. It took several meetings and the involvement of government officials from the district council, including the District Commissioner himself, to convince the community that we could generate our own electricity. We previously had applied to have power from the national grid; we were given a quotation of MK80m to have transmission lines for a distance of 8 kilometres, which we couldn't have managed to raise such an amount; we just withdrew the dream and accepted that we would never have electricity in our community. People walked 8 kilometres to access a maize mill at Mimosa and have their phones charged. Interestingly, very few people owned phones and walked 25 kilometres to Chitakale to access welding services and battery charging, which was time-consuming and costly at the same time.

Village Headman Kalamwa recalls the excitement that inspired the project:

> When work started around 2008, people were excited and knew that this would translate into real development which would transform the community. Everyone, including school-going children, understood that the coming of electricity into Bondo would lead to greater transformation in all sectors such as education, health and the social and economic status as enterprises would blossom in the area.

In every case, the leaders see themselves as helping fulfil a collective dream, with little consideration for any form of opposition or dissent in the village. Electricity comes with its own form of consensus, predicated on ideas of modernity and an ever-changing world of new communication and household appliances. Electricity is presented as a bountiful gift without any costs. Thus, leaders play an important role in adjusting expectations to reality.

Discussions led to the formation of a community management structure known as the Village Electricity Committee (VEC) and developed a constitution to guide their operations. Each village under the project elected a representative in addition to the village chief, also a VEC member. The Village Electricity Committee elects the chairperson, vice chairperson, secretary, vice secretary, treasurer and committee members. The committee consists of a majority of women (more or less 70% of the members), and women have served as chairpersons, too (as of December 2023, the VEC is led by Mrs Delli Nessi, serving her second 3-year term). The VEC also elects members to form the discipline arm, projects arm and business development arm. They recruit and train local villagers as powerhouse operators, transmission power linesmen, and clerks and help develop local leadership skills. Thus, in practice, the VEC serves as a bridge between the community and the operations teams. They handle power supply-related conflicts and customer grievances.

The initial arrangement was that people would access free electricity upon the completion of the project. Later, it was discovered that offering free electricity would be practically impossible because the project's running costs would require finances for paying workers, meeting operational costs, and ensuring coverage for the maintenance and repair of broken-down equipment. A decision was made for household customers to pay a flat rate of MK1000 per month, with chiefs still getting free power supply. This system raised questions of fairness because of the privilege given to chiefs and because a flat rate did not consider different household conditions and varying energy demands. A Business Innovation Facility (BIF) study to develop a sustainable business model for the project recommended the introduction of prepaid metering so that customers pay in relation to their power demand. A further challenge was that MuREA, being a non-profit organization, could not sell electricity to households. This led to the birth of Mulanje Electricity Generation Agency Ltd. MEGA had taken an innovative approach to community-based energy generation: instead of site development followed by 'hand over' to communities, MEGA would continue to run all sites to achieve economies of scale while maintaining the engagement and purchase of the communities. As a social enterprise, MEGA is inclusive throughout its value chain: communities participate in the ownership and governance structures of the organization, in site construction, operation and retail, and form the key target customer group.

10.2.3 Community Energy in Practice

MEGA is a pioneer private energy company in Malawi, generating electricity from 3 powerhouses cascaded along the Lichenya River with a total capacity of 220 kW (60 kW, 110 kW, and 60 kW, respectively). The powerhouses are named after their project period. Bondo 1 got financing from the European Commission in 2012 and later the Tea Estates (mainly Sukambizi Smallholder Tea Association through Fairtrade premiums). Bondo 2 was initially financed internally by MMCT and Practical Action, but the floods of January 2015 washed away the building before the turbine, and the generator was installed. An additional grant from the Colombian Environmental Fund enabled the completion of the project. Bondo 3 got funding from the Scottish Government, the UNDP Global Environmental Facility (GEF), and the Government of Malawi (through the Department of Energy Affairs).

In July 2013, MEGA's first micro-hydro scheme at Lower Bondo on the Lichenya River became partially operational, and the community started seeing results and opportunities immediately. Power was being distributed to households, shops and government health facilities. Businesses were achieving greater turnover due to extended trading hours enabled by the clean lighting. New enterprises were being set up. Bondo's school had nearly doubled the number of teachers on its books. However, the project faced multiple challenges including funding availability (for example the Department of Education has not been able to provide funds to wire the school), enrolment of the community during the operationalization of revenue collection methods, and disruptions in the supply chain.

At the time, in Bondo 1, there was no consideration of standards: people just fell trees and perched them as poles, ready to have their homes connected. One chief, Nkundi, went to source such trees for poles from Lujeri Tea Estates and was given a tractor load. These were enough to connect his whole village. In the meantime, MEGA waited anxiously for a generation licence from the Malawi Energy Regulatory Authority (MERA). When MERA officials visited for assessment and noticed the untreated poles, they advised MEGA to source poles from recommended suppliers. Sukambizi Trust of smallholder tea farmers funded the procurement of treated poles. A Zimbabwe supplier was commissioned to deliver the posts because they provided a longer warranty (35-year, instead of the 12-year warranty offered locally) and the cost, including transportation, was lower than those quoted by local suppliers. However, when the trucks carrying the poles reached the Malawi Boarder Post at Mwanza, the Department of Forestry demanded a certificate that the company had not anticipated. It took the intervention of the Department of Energy Affairs to negotiate the certificate with their forestry counterpart.

The critical challenge facing MEGA now (as of 2023) is to reach scale and financial sustainability while adhering to its founding principles of providing affordable, available, sustainable electricity to 'Bottom of the Pyramid' consumers, reaching everyone in the communities, even the most disadvantaged. MEGA's financial projections predicted operational break-even in Year 6 after the fifth micro-hydro turbine was commissioned. Donor grant funding was needed to build and commission each site, totalling approximately \$2 mn in the first five years and \$1.4 mn for the second five years and over \$1.2 mn has already been spent on the first site and surveys for sites two and three. The upfront infrastructure investment was high and largely charged in US dollars while the potential tariffs are limited by consumers' ability to pay and are largely in Malawian Kwacha.

The availability of local human and technical capacity has also been a challenge. If MEGA is able to achieve and maintain momentum, there could be potential to influence the wider energy landscape in Malawi, paving the way for further investment in private energy provision and facilitating the development of a skills market. MEGA's development has challenged the Malawian administration to review their current licencing frameworks for privately owned energy companies. The new energy bill is likely to make it easier for other privately owned transmission and distribution entities to operate.

The productive use component of the Bondo Micro Hydro Scheme Project works to improve the economic status and livelihoods of the local community in the area through the advancement of economic benefits in utilizing and optimizing opportunities and access to locally generated hydropower electricity in the area. Mega is also strongly interested in increased electricity consumption and revenue from business customers. Hence, MEGA has equally focused on productive uses, social uses, and the institutionalization of the VEC. MEGA lacks an in-depth analysis of community differences and whether any processes of discrimination and exclusion are at work in a society so deeply marked by economic inequality and a complex landscape of post-colonial relations (Fig. 10.3).

The impact of electricity on Bondo's everyday life cannot be overstated. The coming of electricity in the Bondo area has witnessed the mushrooming of different business enterprises, some of which

Fig. 10.3 VEC Installations

being shops with refrigerator groceries items and selling cold drinks, electricity bulbs, freezers (using a home refrigerator to sell freezers/ sweet beer (thobwa); video showing; barber shops and saloons; phone charging; battery charge; tailoring; fresh fish (Kalapao); bars (opaque and bottled beer); burning audio discs; video shooting and editing and many more. Many women have constructed new business models and entrepreneurial activities, such as selling freezers in homesteads, providing them autonomy and empowerment. One middle-aged resident, Line Mangani of Kalamwa Village, says:

> Kale ndikangodikira bambo a m'nyumba andipatse ndalama ndikagule masamba, koma tsopano pano ndikutenga ndalama yanga mkuthamangira pamsika ndekha
>
> (Before starting selling freezers, I used to wait for my husband to give me money to buy vegetables, but now I can decide to go to the market as I do have my own money)

Currently, MEGA has connected over 2000 customers, 13 schools (2 community day secondary schools), 12 maize mills, 1 health centre, and 8 churches and supports over 400 enterprises accessing microfinancing services through a commercial bank to support growing their businesses. The health centre remains MEGA's biggest customer. A big drug storage facility requires power 24/7, and MEGA has been supplying that needed power without interruptions. MEGA still faces a challenge related to how customers can get tokens for the prepaid metres as it is done manually, and it takes three days for one to get the token after paying. Efforts are at an advanced stage to engage mobile money apps where customers will be able to purchase units and input data in their metres while in their homes.

Over the past two years, MEGA has facilitated the introduction of electric cooking to households that are accessing electricity from MEGA. The project fits well with the initial dream of MMCT to have people around Mount Mulanje use the electricity for cooking, thereby displacing the use of biomass. This project is undergoing registration for carbon credits. Households are provided with a double plate hotplate, a wonder box (heat retention fireless cooker), and steel pots for cooking. Each household has a smart metre installed to capture data on household electricity usage and usage for cooking. The pilot had good results related to understanding how much savings a household can make compared to using charcoal or firewood. Initially, 20 households participated in the pilot study, and 100 additional households were issued with the cooking devices. The participating households contributed 30% of the total equipment cost and paid in instalments. Given the success of the projects, current efforts aim at enrolling an additional 500 households to be added to the project in 2023.

10.2.4 A Future Outlook

The Bondo project has made electricity an enabler of social and economic development: reduced poverty and improved well-being, livelihoods and economic development, health and education all have been impacted by local electrification. The diversification of livelihoods also facilitates conservation. New opportunities are opening for the electrification of cooking and the expansion of the network. Learning is accelerating across communities, with the design of new tariffs and prepayment technologies enabling the consolidation of sustainable financial models of management. At the core of the project is the constitution of a social enterprise, MEGA that works as an intermediary between the project and the community. Close coordination with the purpose-organized Village Electricity Committee and tribal chiefs helps guide and prioritize investments and promotes a sense of ownership in the community.

The project, however, has not been without challenges. The aspiration to provide free electricity was soon moderated by the realities and the costs of running a project such as this. MEGA faces even bigger challenges with ambitious plans to continue expanding the project. There is a need, however, for MEGA to explore more closely its means of operation, and the way the community is constituted around it. Are there patterns of difference that MEGA contributes to reproduce, and how? Are there any members of the community who are not yet benefiting from the project? Could MEGA reach them? As challenges are overcome, new questions arise on the horizon.

10.3 Energy Dreams and Energy Futures: An Interview with John Sailence

Chipopoma Community Energy Project: My name is John Sailence, and I come from northern Malawi, from a place called Manchewe. Chipopoma Power is a project that generates electricity for the community using hydropower. We use a 53 kVA alternator, and we generate about 40 kW. So far, we have connected 120 houses and a few businesses around us.

From dreams to reality: The project started... first, it was just like a joke. When I was a kid, I liked to play with some toys that ran on electricity and had batteries, motors, and small lightbulbs. But I lacked electricity, so I started playing with friends who had a better knowledge of electricity. They started introducing me to the type of motors that could produce electricity, like a bicycle dynamo. And we started playing with those things. Then, my mind started growing all kinds of ideas from that bicycle dynamo. Yeah. So, I had a plan that when I grew up, I would have to find a way to generate electricity. I dreamt of using the Manchewe waterfall. Our community is very close to the waterfall, only a few metres. After many years, nobody used this waterfall even though Manchewe is so far from the National Grid. So, I kept dreaming that one day, maybe I would generate electricity to have a barber shop, a TV show, and a phone charger so that the community could work, charge their phone and watch videos here in Manchewe (Fig. 10.4).

Fig. 10.4 John sailence walking towards the manchewe waterfall

Micro-hydropower, in a nutshell: We have a steep place along the waterfall, which is 130 m. So we connected the pipes from the river going down by the waterfall cliff. At the bottom, we built a powerhouse that contains the turbine and the alternator. Then, some of the work went into putting the water in the pipe. That water goes down with high pressure just because it's too steep. At the end of the pipe is a small jet, only 40 millimetres. In this way, the water forces the small jet to get out, with a lot of pressure. Then it hits the turbine, and then the turbine spins. A shaft is attached to the pulley, which is attached to the belt to the pulley of the alternator. So that when the turbine spins, the alternator spins too. And that is how we generate electricity.

Gaining attention: There were a few people who were interested in my ideas, but the first person who ever wanted to talk with me was an entrepreneur called Cameron Mcallister, who used to work at the Mushroom Farm, an alternative lodge in Manchewe. One day, I explained to him my vision of making electricity from the waterfall, and he asked me if that would be possible. I said, with conviction, "Yes!" And many community members were already saying, "Yes, we really need electricity." So, I asked them, "For what do you need it?" And then people answered that they needed it for the barber shop, cooking, and the maize mill. So, I asked myself what kind of device could power a maize mill.

The power of serendipitous encounters: I used to work for The Mushroom Farm as a chef. And that is when I met Cameron and others and started to explain my plans. In this way, I connected with Hastings Mkandawire from Mzuzu, who is building maize mills and other things. Hastings and I worked on a turbine and discussed where we could source an alternator. I returned all excited, and then, back in the community, I met the chef, the chief principal, Gadabassi. I explained my project and my dreams to him and asked for land. It was not difficult to convince him. He said, "No problem, we will help you because we want electricity. We will offer you free land." And that is how we got free land for the project. Now we work in close partnership together in the community. The project is governed by a Board of 13 community members, four men and nice women. As the Board requests, we are always working to empower women and give them a say in the project because women rarely join us in the management of the plant (Fig. 10.5).

Electricity changes communities: There is a very big change. Many businesses developed from the electricity we provide. There are a few barber shops that use electricity, and there are a few salons and a butcher.

Fig. 10.5 The chipopoma power plant still relies on makeshift technology which is improved day on day

And there is the education. If there are any school fails, really, there are just very few: most people here, now they have electricity, get very good results. Maybe some will even go to university just because they had enough time, no, a lot of time, to study their subjects because electricity extended the hours during which they could study. The project has had a very big impact on education and that is really, really good. Now, we are trying to introduce a pump that will be pumping water from the bottom to the hills so that the water goes higher, and then when it runs back, the farmers can divert it for irrigation. This is electricity. It is doing a lot of things. And yes, on top of all that, we have the maize mill! The farmers take maize to process it and make *nsima*.[1] There are a lot of coffee farmers in Livingstonia, and they are using electricity to deshell to remove the skin, which is also having a very big impact on their income.

Transforming cooking practices: People used to cut trees because they needed firewood for cooking. But now, people are increasingly using

[1] A porridge made of maize that Malawians eat for breakfast, lunch and dinner.

hotplates powered by electricity, and they no longer need firewood. And my imagination is now dreaming again. When I see people using electricity for cooking, I think of how I can improve this project so that it has more capacity, enough to support everybody using hotplates and maintain the electricity provision at affordable prices. If you had come to Livingstonia in 2009 and seen how beautiful it was, and then seen it again now, you would think that the land is bare. Before, it was all green. And being green is important to attract tourists and new businesses. People come to Livingstonia for its history because of the Scottish Mission. But the Mission is changing because of deforestation. At Chipopoma Power, we hope to encourage the use of electricity so that people do not cut more firewood.

Connecting people: We started putting the project together in 2017, and in 2022, we started generating electricity for only a few houses. We started with 30 houses, but now we have expanded to 120 houses. We don't have the facilities and resources to expand to more customers. More people are willing to be connected, and we are willing to connect more people all the time. The High Voltage distribution network covers a radius of about 3.5 km, and the Medium Voltage, I think, is now over 12 km, but we keep on adding up: it has grown from 9 to 12 km since we started. The costs of the network were covered by a grant from the UNDP. We also received support from the University of Sheffield, Mzuzu University, and the University of Addis Ababa, who helped the project with phase balancing because some people were consuming a lot of electricity, and this resulted in damage to the generator, which was causing a lot of damage to the generator. CESET visited Chipopoma, and we explained some of the problems that we are facing. I also had some online training, and although it was through the phone, I could understand the problem, and we also got additional technical support to organize the tariff system. This helped the Project to better manage the demand from the project. We invested in switches that could help us to balance the phases and divide electricity equally for people to use electricity according to how much they need to consume because initially, there were many people using hotplates, but the system could not supply sufficient electricity for everyone. The new measures have helped protect the alternator. Without CESET, the project may no longer exist because we would have blown the alternator.

Networks of power and networks of people: Connecting with universities helps us a lot. Different universities, sometimes from other countries, have visited the Chipopoma Power Project. The project CESET, for example, brought different people from different universities and countries. Then, I could express myself and the problems that I'm facing. They were able to work with me to find out how to solve these challenges and help me access the technical expertise that I needed, from understanding the problem to defining measures for the phase balance to improving my calculations about the network load. And the challenges are multiple. CESET also supported us in buying protection from lightning, just because Livingstonia is at a very high altitude and we have many electrical storms. One storm blew one of the transformers. This could also cause damage to the houses. Those protections were very important to keep the project going (Fig. 10.6).

Big dreams, big hopes: I am always planning for Chipopoma to grow. We have discovered that we have the potential to generate more electricity. In five years, I would like to generate more electricity so that we can have a lot more customers as well as to sell some of the electricity to the national grid. Currently, the Chipopoma Power Project is not sustainable because we have very few customers. We depend on grants and ad hoc support, such as that provided by CESET. I would like Chipopoma to become a good example for our country.

The future of community energy in Malawi: Community energy is only starting in Malawi. It is very slow, and very few people can access this. I would like to spread this model and explain our experiences so that everybody has access to this electricity because as of now, very few people have electricity in the country. There is a lot of demand. There are still areas we have to reach that we have not reached yet because we do not have enough resources to extend the network. But the demand is high and growing.

10.4 Conclusion: Community Energy in Practice

The experiences of Bondo and Manchewe demonstrate the important role that community energy can play in accelerating the transition to sustainable energy in Malawi. However, they do not provide a triumphalist account of their projects, and they present them as work in progress not only because they have plans to extend the network but also because the continued learning involved in the project means that a community energy project is always a work in progress. Though both projects share many characteristics (e.g., a mountainous environment, the use of

Fig. 10.6 The manchewe waterfall continues to inspire

hydropower, and a similar regulatory environment), the projects present important differences. Talking with each other, Arnold and John joke that they are at different stages of development.

One of the main differences, however, is that the projects have been developed in extremely different institutional environments. MEGA emerges from a concern with conservation, and the relationships with the Mulanje Mountain Conservation Trust and Practical Action have enabled the ongoing continuation of the project, even when facing difficult challenges. Moreover, their institutional collaboration has facilitated further support from other institutions, such as the University of Strathclyde. In contrast, the project of Chipopoma has not had that continuous support and has relied on the personal networks of John Sailence himself and local businesses such as the Mushroom Lodge. While John's heroic story of an individual fighting for a dream constitutes a good storyline, it comes at a significant personal cost and may hinder the long-term sustainability of the project. The long-term sustainability of the Chipopoma Power Project depends both on extending the network of customers to make it financially viable and extending the institutional network of institutional partners so that ad hoc support can be found when needed.

This is the reason why sustainable energy scholars in Malawi have advocated for the government to provide institutional and financial resources to small-scale projects that as the Chipopoma Power Project (see Chapter 9), to harness the potential of each community to take control of their energy futures.

CHAPTER 11

Bringing Policy to Light: Implementation Challenges for Mozambique's Off-Grid Policy Regulation

Lorraine Howe, Carlos Shenga, and Carlos Alberto Cuvilas

Abstract In 2021, Mozambique approved a new regulation for access to energy in off-grid areas (Decree 93/2021) in December 2021. This chapter explores how the implementation of the regulation creates opportunities and challenges for the uptake of off-grid energy projects in Mozambique and the extent to which the new energy policy contexts facilitate the development of community energy projects.

The chapter analysis combines a policy analysis of the regulatory landscape and key informant interviews with energy operators conducted in 2022 in the wake of the adoption of the new regulation. The regulation

L. Howe · C. Shenga (✉)
Centre for Research on Governance and Development, Maputo, Mozambique
e-mail: cshenga@cpgd.org.mz

L. Howe
e-mail: lhowe@cpgd.org.mz

C. A. Cuvilas
Faculty of Engineering, University Eduardo Mondlane, Maputo, Mozambique
e-mail: carloscuvilas@gmail.com

V. Castán Broto (ed.), *Community Energy and Sustainable Energy Transitions*, https://doi.org/10.1007/978-3-031-57938-7_11

sets the foundation for off-grid energy uptake through increased confidence and investment in both the off-grid energy market and government institutions. However, challenges remain, such as the transparency of roles and responsibilities of key government institutions and having appropriate resource mechanisms in place—e.g., financial incentives or human capital, to facilitate the regulation's implementation.

Keywords Mozambique · Off-grid energy · Regulation · Energy transition · Implementation

11.1 Introduction

In December 2023, amidst the fanfare of the annual climate change summit or Conference of Parties in Dubai (COP28), President Filipe Nyusi launched Mozambique's Energy Transition Strategy for 2023–2030. He summarised his declarations in a tweet saying:

> Compatriotas,
>
> Enfrentamos duas realidades que constituem dilema face às nossas ambições de desenvolvimento económico: como um país em desenvolvimento, os níveis de acesso à energia estão abaixo de 53 por cento, mas somos hospedeiros de grandes reservas de gás natural.
>
> Para contornar esta situação, acabamos de lançar, em evento paralelo à COP28, a Estratégia Nacional de Transição Energética, assumindo desta forma o nosso papel fundamental na transição energética no suprimento das necessidades energéticas da África Austral, o que vai contribuir para a descarbonização regional.
>
> Esta iniciativa não apenas coloca Moçambique na vanguarda da inovação climática, como também o posiciona como um destino de investimento atractivo sustentável.[1]

[1] "Compatriots,

We face two realities that constitute a dilemma in the face of our economic development ambitions: as a developing country, energy access levels are below 53 per cent, but we are home to large reserves of natural gas.

To overcome this situation, we have just launched, in a parallel event to COP28, the National Energy Transition Strategy, thus assuming our fundamental role in the energy transition in supplying Southern Africa's energy needs, which will contribute to regional decarbonization.

Nyusi's declaration highlights the country's priorities for the energy transition: exploiting the large deposits of natural gas found in the Rovuma Basin in 2012 (whose reserves are estimated at 286×10^9 m^3) and attracting international investment. The political declaration aligns with recent interventions by the African Union, which in June 2022 called for African countries to use their "abundant energy resources, including renewable and non-renewable energy, *to address energy demand*" (African Union, 2022). Yet, Mozambique's Transition Strategy will direct an investment of USD 80 billion to renewable energy to reduce the country's dependence on fossil fuels. The Strategy adds to a series of reforms in the energy sector, including a new regulatory framework for off-grid developments, that demonstrates an expansion of the role of renewable energy in the future of the country.

The dynamic changes in energy policy witnessed in Mozambique in recent years follow the Government's ambitious commitment to achieving universal energy access by 2030. Since 2001 the electrification rate has increased from 5% to an estimated 51.3% at the end of 2023 (EDM, 2022). Despite these gains, a lot remains to be done to achieve universal access, particularly for Mozambique's rural communities, which comprise about 67% of the population (INE, 2019p. 38). Less than 5% of rural dwellers have access to electricity compared to 57% of urban dwellers (INE, 2022). To attain the 2030 target, a combination of national grid expansion and off-grid electrification through mini-grids and solar home systems is planned (Government of Mozambique, 2018). Off-grid expansion is crucial for reaching rural communities, which are widely distributed across the country, often in low densities, making it financially and logistically difficult to instal, supply, and maintain electricity infrastructure.

To help increase Mozambique's energy access, regulatory reforms have been made to enable the private sector to supplement the energy provided by the state utility company (Electricidade de Moçambique, EDM). The first reforms in 2012 and 2013 approved regulations to enable Independent Power Producers (IPPs) of large (Decree 16/2012) and small (Decree 69/2013) projects to produce and supply energy to the utility company grid system. The second, in 2021, was to regulate energy access in off-grid areas for public and private entities through mini-grids up to

This initiative not only places Mozambique at the forefront of climate innovation but also positions it as an attractive, sustainable investment destination".

10 MW and energy services (Decree 93/2021). The regulatory framework for off-grid energy access was further developed during 2021/22 by the National Regulatory Authority (ARENE) in partnership with BRILHO, an FCDO/Sida-funded programme.

A transition, however, does not end with the approval of regulations. What is the impact of these policy changes on the energy systems in Mozambique, particularly regarding the government's objective of achieving universal energy access by 2030?

The development of the off-grid regulation during the lifetime of the CESET project provides an opportunity to understand how this regulation is impacting Mozambique's off-grid energy landscape in broad terms, specifically on community energy uptake. How the regulation will be implemented and what projects and actions will follow from the regulation is crucial to understanding the extent to which the off-grid energy regulation will indeed facilitate a transition and, in particular, support the development of community energy. An analysis of the existing policy landscape is combined with a study of key stakeholders working in the country's off-grid energy sector in terms of the opportunities, gaps, and challenges they perceive regarding implementation.

The following section (Sect. 11.2) provides a brief overview of the theoretical basis of the study, based on implementation theory, and informs the methods, presented in Sect. 11.3. The empirical results are presented in two sections: Sect. 11.4, which presents the policy analysis, and Sect. 11.5, which presents the perspectives of policymakers captured in qualitative research. Section 7 discusses the findings of these two sections, concluding that while setting a positive regulatory environment, there are still challenges in the implementation of off-grid energy projects.

11.2 From Policy to Implementation

After a complex process of adoption of the off-grid regulation, the question is what it means in practice. The literature on public policy and administration often assumes that there is a divide between a process of policymaking involving bureaucrats and a process of implementation, in which different 'active' actors adopt and work on the policy. Implementation is considered a late stage in the process, a residual process to achieve the policy goals (Hupe & Hill, 2016).

A common approach is to think of 'policy' as the core that generates a series of actions during implementation. The policy is thus defined as prescriptive, while implementation consists of applying those prescriptions. Implementation actions thus may involve the further development of policy objectives, the adoption of instruments that provide precise detail about how to make the policy effective, as well as the specification of political-administrative arrangements to organise institutional actions, and any procedural rules (Knoepfel et al., 2007). However, the relationship between policy and implementation is not straightforward. There is a burgeoning literature that explores the complexities of this relationship, particularly questioning the separation of co-jointed processes of policy formation and policy implementation. This kind of thinking may easily lead to a conceptualisation of implementation as 'the rest', the kind of actions that follow a complex strategic process of policy making (Hupe & Hill, 2016). In this chapter, the off-grid regulation is approached as part of an ongoing process of change, shifting towards the greater opening in the energy policy of Mozambique and influenced by liberalisation trends that have dominated infrastructure thinking since the 1990s.

This approach intends to overcome stylised conceptualisations that assume the strict separation between policy and implementation and, in doing so, overlook the struggles and compromises involved not only in policy formulation but also in its interpretation. For example, empirical experiences of delivering policy to achieve transitions to sustainability show that implementation raises specific challenges, often determining the success of a policy, for example, in the integration of new technologies within existing infrastructure systems (Aguiléra & Boutueil, 2019). Implementation is closely related to the strengthening of networks and alliances and thus, who is represented in the implementation process will determine the extent to which its outcomes are just and fair. The incorporation of communities in the co-production of implementation processes is increasingly a demand for sustainable development (Diep et al., 2022).

So far, the literature examining the development of off-grid policy, especially in African countries, has tended to pay more attention to the development of regulatory frameworks alone without fully appreciating the complexities of implementation (e.g., Bhattacharyya, 2013; Ole, 2020). We have identified five crucial elements that facilitate the advancement of off-grid policy towards implementation: clarifying institutional roles, enrolling different actors including the private sector, attracting resources and finance, addressing material constraints such as supply chains and facilitating the involvement of the community.

11.3 Methodology

The focus of the research was to consider the institutional context of policy development and how it will affect the implementation of off-grid regulation in Mozambique. Following the discussion above, the first step was to situate the off-grid regulation within the wider landscape of energy policy in Mozambique. This involved a systematic review of existing legislation exploring how off-grid, and more particularly community energy, is supported or restrained by the policy landscape. This review also enables mapping the actors active in the energy landscape, helping explore what different roles actors play.

The second part of the study consisted of eight in-depth key informant qualitative interviews with energy operators, bilateral and multi-lateral donors, and organisations working in Mozambique's off-grid energy sector. This is a small sample, but it represents a relatively small group of stakeholders who are actively involved in embedding the current off-grid regulation in the energy sector. Interviews were conducted in August and September 2022, using a semi-structured interview guide with questions on the content of the regulation, the process that led to the development of the regulation, key challenges and how implementation of the regulation could be monitored going forward.

The interviews were transcribed and anonymized, and the data was analysed using Nvivo software. Additional feedback and comments were sought from key stakeholders to ground truth and validate the draft findings before they were finalised. A first round of analysis focused on understanding the consequences of the regulation during its implementation, including opportunities for off-grid energy expansion, the emerging challenges for off-grid development, and the possibilities to develop community energy as a special case of bottom-up project development. The policy recommendations emerging from this analysis were compiled in a CESET Briefing (Shenga & Howe, 2022). Further analysis examining the role of off-grid energy in the current energy landscape of Mozambique, the changing roles of actors, and the resources available for off-grid energy development informed the discussion.

11.4 The Development of Off-Grid Policy in Mozambique

Electrification is central to the delivery of the SDG7 in Mozambique. The emphasis has gone into extending the central grid to rural and remote areas traditionally excluded from having energy access. Despite appreciable growth in the rural electrification process, according to forecasts, Mozambique's current rate of electrification would only achieve global access in 2065. According to the World Bank, at the current rate, "Mozambique will connect 6.2 million users, reaching 63–65 per cent of its population by 2030, through grid connections" (World Bank Group, 2024), which is positive but far from the optimistic government commitment of reaching universal electrification by 2030. The World Bank recommends diversifying the range of on-grid and off-grid responses to bridge this gap, particularly considering that the increasing risk of drought in the country is likely to compromise its electricity system dependent on hydropower.

The government leads electrification efforts. The Ministry of Mineral Resources and Energy (MIREME) is the main energy planning entity in Mozambique, which has attributed competencies to the exploitation of energy resources and the expansion of energy infrastructure (MIREME, 2021). However, MIREME operates in coordination with the Ministry of Economy and Finance (MEF), which allocates funds for infrastructure expansion, and with Electricidade de Moçambique, E.P. (EDM) the public enterprise that manages the grid and implements electrification projects. While EDM implements on-grid projects, the Fundo de Energia (Energy Fund, FUNAE) is the agency responsible for the implementation of off-grid energy projects such as Solar Home Systems and mini-grids. FUNAE was created for the promotion of renewables in 1997, but it was restructured in 2020 (Decree 101/2020, of 12 November) to facilitate its role in the financing and implementation of off-grid energy projects. The project of electrification has a strong territorial component, with EDM being perceived as the main entity implementing projects in urban, high-density areas and FUNAE being responsible for rural, remote, and low-density areas. This separation means that the potential for off-grid projects in higher-density areas in Mozambique remains unexplored.

The main legal instrument to facilitate electrification is the Energy Law adopted in 2022 (Law No. 12/2022) which in contrast with the previous Electricity Law of 1997, pays special attention to facilitating the adoption and integration of renewables. The 2022 Electricity Law complements the Power Infrastructures Master Plan 2018–2043 aims to facilitate the development of new renewables for their integration into the national grid, fostering some important solar projects such as the Mocuba Solar Plant (30 MW of installed capacity). Regarding increasing off-grid capacity, in 2022, FUNAE had 97 functional solar plants across the country (ALER/AMER, 2022). Throughout 2023, three more mini-grids came into operation in the southern part of the country, totalling 100 (FUNAE, 2023). and adding 9.4 MW of installed capacity (see also additional data at ALER/AMER, 2023). Despite this progress, solar and wind energy still make a limited contribution to the government's electrification efforts. Much of this progress depends on the support (financial and technical of international cooperation partners, such as the European Union and UNIDO or bilateral agreements).

However, since the 1990s, several policy strategies around energy have been directed towards attracting private companies to the energy sector. The Electricity Law of 1997 first introduced Power Purchase Agreements (PPAs) to facilitate greater participation of private companies, particularly for electricity generation. Over decades, the role of the private sector has diversified, and increasingly it is being perceived as playing a key role in the development of renewables. For example, the Power Infrastructures Master Plan 2018–2043 envisages to the grid adding 125 MWof solar generation capacity, developed through Public Private Partnerships. In 2020, the government partnered with the European Union to develop a Renewable Energy Auctions Programme to facilitate a competitive and transparent procurement process for renewable energy and to reassure IPPs of the procedures for the development of new infrastructures, providing a more secure investment landscape. The Autoridade Reguladora de Energia (Energy Regulation Authority, ARENE) is the authority that oversees competition between public and private operators. ARENE was established in 2017 (Law n. 11/2017) as an independent, autonomous body, which conferred it a greater degree of legitimacy than its predecessor, the governmental body Conselho Nacional de Electricidade (National Electricity Council, CNELEC).

A report by the Lusophone Association for Renewables (ALER) and Mozambique's Association for Renewables (AMER) highlights the important role that Independent Power Producers (IPPs) can play in developing the landscape of off-grid energy in Mozambique (ALER/AMER, 2022). Here is where the off-grid energy regulation provides an important framework to fill what ALER and AMER call 'a legal void' in defining what off-grid areas are and what the procedures are to obtain an off-grid concession. The new off-grid regulation may facilitate the active participation of the private sector in the development of mini-grids (currently dominated by FUNAE).

Optimistic assessments of community action on energy do not translate easily into context in which government-driven public interests wrestle with the potential development and economic opportunities of the energy sector Bomberg and McEwen (2012). In Mozambique, both public and private providers faced the regulatory requirement to demonstrate social benefit from any infrastructure or economic development project. Engaging communities is often thought to facilitate decentralised supplies and the adoption of heterogeneous renewable generation technologies (Wirth, 2014). However, there are broader questions of technology justice and resource sovereignty that remain unaddressed in technological approaches to community energy. Often, inadequate legal frameworks, lack of access to technology, and lack of resources combine to prevent communities from instituting, managing, and owning energy projects (Ambole et al., 2021). Mozambique is not an exception. Policies for the institutional diversification of the energy sector have put greater emphasis on infrastructure development and increasing investment than on ensuring social development, despite increasing realisation that a transition to sustainable energy depends on facilitating a fundamental cultural and social change. In fact, the discourse on transition appears to have fostered a backward movement away from the participatory approaches that feature in previous policies. For example, the 2009 New and Renewable Energy Development Policy (No. 62/2009) stated its intent to "*Promote broad participation of the community, companies, civil society, in the process of developing new and renewable energies to guarantee the equitable prioritization*". In contrast, the 2018 Electricity Infrastructure Integrated Master Plan 2018–2043 (MMRE, 2018) makes no reference to community participation or to community energy. The off-grid regulation emphasises community engagement as a condition for the development of mini-grids but does not specifically state community energy as a workable strategy for development. In between the public and the private sector, communities and civil society organisations get squeezed out. So far, there is limited evidence of community engagement in the 100 mini-grid projects led by FUNAE, let alone facilitating their leadership in the development of solar projects.

11.5 Policymakers' Perspectives on the New Off-Grid Policy Regulation

Against this context, policymakers' perspectives offer a contrasting outlook of optimism and scepticism about the possibility of delivering a renewable, off-grid future. Scepticism increases in relation to community energy, which some actors do not see as a workable alternative. The analysis explored first the positive aspects of the off-grid policy, then the perceived implementation challenges, and finally, explored community energy perceptions in the context of the new policy.

11.5.1 Positive Aspects of the New Policy

All key informants emphasised the positive significance of having an off-grid regulation approved and in place. The regulation is unanimously seen as a key milestone for the development of the energy sector in Mozambique and particularly for achieving universal energy access in rural and hard-to-reach communities.

While there was acknowledgement that the regulation may not meet everyone's requirements, there was widespread agreement that simply having a regulation in place, even if not perfect, is preferential to having nothing at all. This is because the 'rules of the game' of the operating environment are now known and publicly available. This has increased the private sector and investor confidence in the off-grid sector previously monopolised by FUNAE.

The interviewees stressed that the increased confidence in the business operating environment presents several opportunities. Firstly, there is an expectation that the off-grid market will develop by becoming more competitive, with the entrance of new companies and private investors, particularly in relation to mini-grids. Up to now it has been very challenging for mini-grid operators to advance in the sector given the grey areas that existed around licencing, compensation arrangements and the national electrification expansion strategy, for example. Secondly, linked with the expected off-grid market development is the anticipation of accelerated financing. This is expected from both private sources and development assistance providing much-needed startup capital or investment guarantees. However, as some respondents noted, it is a challenging business environment at the moment. Local energy operators are having to cope with multiple shocks from COVID-19, the Russia-Ukraine conflict, cyclones and a heated conflict in the north of the country. Moreover, local operators may not have the capacity to take advantage of the financing available. For example, they may not meet the eligibility requirements to access the funding or have the skills to apply for or absorb the investment.

Energy operators, donors and implementing partners also speak of their increased confidence in the government and its commitment to facilitate a conducive operating environment for the off-grid sector. One further opportunity identified is that having the off-grid regulation approved and in place could facilitate financial incentives such as tax breaks, subsidies, and tariff support to make the financial returns on investment more attractive for potential off-grid energy suppliers. In August 2022, Mozambique's President Nyusi announced a stimulus package for the acceleration of the economy (Pacote de Aceleração Económica, PAE) to respond to the country's growth needs. In addition to reducing the VAT rate from 17 to 16%, one of the measures announced—a VAT exemption on imports of electrical items—was specifically targeted at promoting investment in renewable energy and expanding energy access in rural areas. As of February 2023, specific details related to this announcement, such as exactly what electrical items are covered by the exemption, are still to be disclosed. The fiscal authority (Autoridade Tributária de Moçambique, AT) and the Ministry of Economy and Finance (Ministro da Economia e Finanças, MEF) have received from MIREME and from a representation of national off-grid energy operators a detailed proposal on the interpretation of PAE for the off-grid sector, and a list of materials and equipment for which custom tariffs and vat exemptions would be applicable on imports.

The government is seen as taking a proactive approach to developing legal and regulatory reforms. This extends beyond developing and approving the off-grid decree but also in the revisions to the electricity law and, most notably, institutional development through the creation of ARENE as an independent authority. There was widespread praise for the consultation process that led to the development of the off-grid regulation, both by those who participated in it as well as those who observed the process indirectly. Both ARENE and MIREME are seen as having played a positive role in developing a supportive policy environment as both organisations have engaged in constructive discussions with stakeholders working in the sector. Most interviewees were encouraged by this in the context of ARENE being an emerging institution that is still learning and building its capacity and whose remit covers the full spectrum of the energy sector, of which off-grid is only a small component. Expectations are, therefore, high among those interviewed for this conducive relationship to continue. It has also led to hopes for additional institutional reforms, specifically those that could pave the way to clarify the roles and responsibilities of EDM and FUNAE in how they work together with the private sector for off-grid sector development.

11.5.2 Perceived Implementation Challenges

Despite the overall optimism and opportunities conveyed regarding the off-grid regulation, respondents expressed reservations regarding its implementation. At the time of the interviews, a suite of accessory regulations (27 instruments in total) needed to be finalised and/or approved and published by the government, covering a wide range of topics such as concessions and technical and safety standards. Against this backdrop, one of the main challenges expressed is that the regulation has not yet been tested in practice, and there are several areas of ambiguity outlined below, which it is hoped will be clarified by the approval of the accessory regulatory instruments.

The off-grid regulation does not make any reference to the role of key public institutions involved in and responsible for the energy sector. Specifically, the lack of reference to the roles and interaction between EDM, which in cooperation with MIREME is responsible for the national electrification strategy, and FUNAE, which is responsible for energy access projects in rural and off-grid areas, was seen as a challenge for many respondents. The opaque nature of the planning and decision-making at the central level on the national electrification plans can interrupt investment in off-grid projects. This is because there is no clarity as to what happens when off-grid and on-grid energy projects interact or if grid expansion takes place in an area allocated to off-grid energy projects.

Another major challenge foreseen by respondents relates to financing and tax incentives. There was widespread anticipation that the decree would provide information about the fiscal arrangements for off-grid investments, such as taxation, customs duties, tariffs, subsidies, and compensation arrangements in the event the national grid extends into off-grid concession areas. However, the decree does not provide information on these issues. A concerted effort led by AMER brought a package of incentives for approval by the government, but the final approved incentives were only a fraction of those proposed.

Mozambique represents a difficult business operating environment for off-grid energy projects, and it is critical that operators and investors know the financial viability of their projects before they can commence operations (see also Chapter 5). This was emphasised most strongly by mini-grid developers. This is due to the time and sunk costs of conducting feasibility studies, developing a workable business model within the existing tariff regime and the exposure of capex costs. Similarly, a respondent who operates a solar home system (SHS) business, explained that 45% of operating costs are related to customs duties and tax to import the SHS equipment. These costs must be passed directly onto the consumer in the prices charged. Yet, the target consumers of SHSs in off-grid, rural areas are typically people with little or no regular income or ability to pay for goods and services. A reduction or exemption of import duties, for example, would make these systems much more affordable and would increase their uptake.

Another challenge foreseen, although not solely related to the off-grid decree, is whether there are sufficient government resources (financing and human capital) and alignment to implement the decree and the wider, ambitious universal energy access target. Respondents acknowledged that while the off-grid energy sector has gained momentum there remain skills and capacity gaps to implement and manage the new developments. The energy transition plan expects to create the conditions for a USD80 billion investment, but the reliance on international development partners is still high.

Other challenges are institutional. ARENE is seen as being stretched to respond at the pace of the off-grid sector due to it being a new institution that is not yet operating at full staffing capacity. Other public institutions are seen to have related parts to play, such as MIREME, EDM, FUNAE, and others, such as the National Institute for Standardization and Quality (Instituto Nacional de Normalização e Qualidade, INNOQ) which has responsibility for enforcing quality standards for off-grid energy equipment It was felt too early at this stage to know whether these institutions have clarity in terms of their roles in relation to the off-grid decree and have the capacity to deliver and coordinate their work during its implementation. Energy operators, donors and implementing organisations expressed concern that if the government is too slow to act because of resource constraints, financial and technological opportunities could be missed. The dynamic nature of the renewable energy sector could make current policies and practices quickly outdated.

In fact, some operators already identified some missed opportunities from a technology point of view, particularly in delivering small—and micro-scale systems such as SHSs. Since the off-grid regulation was designed specifically to accelerate the development of mini-grids in the country, other types of off-grid technology risked becoming marginal, regardless of their potential. While the regulation classifies systems into three mini-grid categories, smaller-scale systems are still subject to relatively large amounts of reporting requirements, which has financial and operational implications for the operators. The failure of the regulation to embrace a wider range of technology options, particularly those aimed at the low end of the market, was felt by some providers as a missed opportunity.

Finally, an indirect challenge emerged during the discussions related to overseas development assistance. While not specifically linked to the regulation itself, this challenge relates to the off-grid energy sector in general and was referenced in at least three interviews. Respondents spoke about the lack of coordination among donor organisations that are providing development assistance to support the off-grid energy sector. There is a vibrant community of donors working in the sector providing essential financing and technical support. However, there was a degree of frustration about the perceived homogeneity of donor programmes and the extent to which the off-grid regulation actually challenges such models. Interviewees called for a more strategic, coordinated approach where donors supported a wider range of actions to target 'blockages' in the sector rather than just providing catalytic grants and results-based financing.

11.5.3 Potential for Community Energy Uptake

Interviewees were asked specifically to evaluate the off-grid regulation in relation to its potential to promote community energy. One respondent emphasised that the first and foremost guiding principle, stated in Article 2 of the decree, is that all projects that involve activities to supply access to energy in areas outside of the national grid are "in the public interest and of a social nature". In their view, having this definition at the outset of the regulation provides a clear signal that the intention of the regulation, at its core, is to provide social (and economic) benefits to rural communities.

Several respondents expressed cautious optimism, that there is an underlying assumption that this regulation will facilitate energy access in rural areas through community energy projects. However, it was noted that this has not yet been tested in practice.

At the same time, there was consensus that community energy is not well highlighted in the regulation, and it is not clear how communities can or will be involved or participate in off-grid projects. One interviewee, for example, noted community energy is not salient in the context of the 'market-based' approach that seems to be dominating the off-grid energy sector. In their opinion, this is often top-down from the supply side with limited community participation, or where there is community participation, this is often focused on awareness raising rather than on community needs.

The challenge of financial incentives, raised in the previous section, was also noted as critical in enabling the uptake of community energy projects. Unless import duties and taxes are addressed to reduce tariffs, community energy projects, especially those targeting the most vulnerable, will be very difficult to pursue. According to those interviewed, the sustainability and financial viability of many off-grid energy operators are on a knife edge as they try to recover from the economic shocks of the COVID-19 pandemic and rising inflation. In turn, their ability to absorb any further shocks from the market is limited, and they need a stable operating environment to survive. This was emphasised in relation to the micro-scale energy operators whose main clients are those most affected by energy poverty. The rising costs of living due to inflation mean that not only is the ability of people to pay for energy services even further reduced, but the energy operators (without any financial incentives) have to pass their rising costs onto the consumer since their financial returns on investment are already stretched to the limits. To adapt to these circumstances, some energy operators reported they are trying to focus more on productive use systems so that instead of just providing energy access, they can also help their customers generate income through activities that might include irrigation, refrigeration, or milling, as a potential win–win solution for both parties.

Reflecting on the wider implications, the likely consequence noted by all respondents is that energy operators will prioritise those communities where there is a higher ability to pay. Those communities that have potential economic development or anchor clients that can provide stable revenues will be targeted first over more marginalised communities and will be left behind to benefit last. Among interviewees, there was little sense of facilitating greater autonomy, and generally, community energy was reduced to 'benefits to communities' rather than 'led by communities.'

11.6 Discussion and Policy Recommendations

Several highlights emerge from the perspectives of energy operators, donors, and implementing organisations working in Mozambique's off-grid sector regarding the opportunities and challenges presented by the new off-grid regulation. Table 11.1 summarises the insights from the analysis above.

Reading the insights above, considering the discussion on implementation, the new off-grid regulation must be understood as a building block in a broader process of policy development that also involves action on the ground. Respondents emphasise that an existing regulation is better than a perfected future one because it provides the opportunity to test its principles in practice. While the extensive consultation had also helped to enlist the support of diverse stakeholders, all of them show legislative change as a milestone rather than an endpoint. At the time of the interviews, much hope was put in intermediary policy instruments, such as the additional 27 accessory instruments that complete the off-grid framework to provide prescriptions for action, but even as the details of those instruments become known in early 2023, uncertainty remains in a dynamic process in which the interaction between policy and technology is constantly shifting.

One priority for the energy sector is the provision of appropriate financial incentives for growth in the off-grid sector. Energy suppliers are operating under difficult economic circumstances and in a country context where all equipment for off-grid energy systems must be imported because there is no local production capacity. Appropriate tax breaks and other financial incentives are crucial for the off-grid energy sector in general and for community energy in particular. In the context of communities with very low incomes, the success of community energy depends on having low tariffs. However, having little empirical evidence of the delivery of community energy in Mozambique, cost limitations seem insurmountable without government support.

Table 11.1 Implementation dimensions

Aspects of the successful implementation of off-grid projects	*Contributions to theoff-grid regulation*	*Perception of policymakers*	*Consequences for community energy*
Institutional leadership and coordination	Increase clarity in institutional roles An addition to a changing political economy of energy, challenging the territorial assumptions of previous policies	Consolidation of governmental institutions within the energy sector, and development of partnership approaches across the sector	Lack of acknowledgement of the role of community energy in delivering the transition to sustainable energy but opening of spaces of experimentation within the energy infrastructure landscape
Inclusion of the private sector	Clear frameworks and procedures for the private sector's participation in the off-grid sector	An intermediary step in a changing trajectory to facilitate the inclusion of the private sector	Opportunities for the development of community energy projects led or in partnership with new private operators
Attracting finance	Increasing investors' confidence in the sector	Need to overcome barriers related to the returns of investment in a complex market environment	Need to understand how new opportunities can be harnessed by communities through small, localised projects

(continued)

Table 11.1 (continued)

Aspects of the successful implementation of off-grid projects	*Contributions to theoff-grid regulation*	*Perception of policymakers*	*Consequences for community energy*
Overcoming material constraints	Access to affordable supply chains	Need to understand the wide variety of off-grid projects that could take place beyond large-scale, private operated mini-grids Need for ensuring VAT discounts and tariffs subsidies to access imported technologies and components	Heterogeneous generation projects operating at different scales and an affordable supply chain may facilitate the development of tariff models for the most disadvantaged communities
Community engagement	Explicit statement of the need for off-grid projects to benefit communities	Prioritisation of communities with the ability to pay over those who are disadvantaged Limited concern for community skills and leadership	Need to claim active enrolment of communities in mini-grids and other off-grid projects

The regulatory reforms have created momentum and raised expectations in off-grid sector development. Some of those relate to the perception that more clarity is needed in the energy sector. The regulation has a performative role in creating the context of implementation, facilitating institutional development, and creating the conditions for actors to take action on the ground. Ongoing consultation further facilitates implementation. However, there are unrealistic expectations about the extent to which regulation alone can bridge the gap of capacity and skills faced by all the organisations involved in the regulation, from ARENE to every small operator.

However, the development of regulation creates capacity, by situating organisations in the institutional landscape. For example, the definition of FUNAE and EDM's remit and the arrangements should be off-grid and on-grid projects meet the explicit expectation placed on ARENE in fulfilling its mandate to regulate the sector and help orient other actors within the sector. The off-grid sector is still nascent and the absorptive capacity of energy operators, whether they are private businesses, associations, or community groups, requires time to develop.

As Table 11.1 shows, the off-grid regulation both supports and constrains community energy. The off-grid regulation does not have a strong emphasis on community energy. Article 3 of the regulation specifies that it applies to "collective entities of private and public realm", which does not address community energy explicitly. In practice, community energy may be already happening. Implementing partners such as UNIDO, in partnership with FUNAE, have been training community associations on how to apply for off-grid community energy funds disbursed by BCI Commercial Bank (UNIDO, 2022). However, despite these experiences, it will require time to monitor the number and type of community energy systems that may be established under the new regulation before an assessment can be made regarding the regulation's impact in this area.

References

African Union. (2022). *African energy transition programme*. Retrieved January 16, 2024, from https://au-afrec.org/energy-transition-programme

Aguiléra, A., & Boutueil, V. (2019). Chapter 4—Implications for public policy. In A. Aguilera & V. Boutueil (Eds.), *Urban mobility and the smartphone* (pp. 143–167). Elsevier. https://doi.org/10.1016/B978-0-12-812647-9.00004-4

ALER/AMER. (2022). *Renewables in Mozambique*. Retrieved January 16, 2024, from https://www.lerenovaveis.org/contents/lerpublication/resumo-renmoz2023.pdf

ALER/AMER. (2023). *Renewables in Mozambique*. Retrieved January 16, 2024, from https://www.lerenovaveis.org/contents/lerpublication/resumo-renmoz2023.pdf

Ambole, A., Koranteng, K., Njoroge, P., & Luhangala, D. L. A. (2021). Review of energy communities in sub-saharan Africa as a transition pathway to energy democracy. *Sustainability*, *13*, 2128. https://doi.org/10.3390/su13042128

Bhattacharyya, S. C. (2013). To regulate or not to regulate off-grid electricity access in developing countries. *Energy Policy*, *63*, 494–503.

Bomberg, E., & McEwen, N. (2012). Mobilizing community energy. *Energy Policy*, *51*, 435–444. https://doi.org/10.1016/j.enpol.2012.08.045

Diep, L., Mulligan, J., Oloo, M., Guthmann, L., Raido, M., Ndezi, T., Diep, L., Mulligan, J., Oloo, M. A., Guthmann, L., Raido, M., & Ndezi, T. (2022).

Co-building trust in urban nature: Learning from participatory design and construction of nature-based solutions in informal settlements in East Africa. *Frontiers in sustainable cities*, *4*. https://doi.org/10.3389/frsc.2022.927723

EDM. (2022). *The status of renewable energies and the role of EDM in the energy transition*. Presentation delivered by EDM representative (Directora Olga Madeira Utchavo) on 28 November 2022 at CESET Workshop.

FUNAE. (2023). *Noticias*. Retrieved January 16, 2024, from https://funae.co.mz/noticias/

Government of Mozambique. (2012). *Boletim da Republica. Decreto n. 16/2012, 4 Julho*. Regulamento da Lei sobre Parcerias Público-Privadas, Projectos de Grande Dimensão e Concessões Empresariais.

Government of Mozambique. (2013). *Boletim da Republica. Decreto n. 69/2013, 20 Dezembro*. Regulamento de Parcerias Público-Privadas e Concessões Empresariais, de Pequena Dimensão.

Government of Mozambique. (2018). *Estratégia Nacional De Electrificação 2018–2023*.

Government of Mozambique. (2021). *Boletim da Republica. Decreto n. 93/2021, 10 Dezembro*. Regulamento de Acesso à Energia nas Zonas Fora da Rede

Government of Mozambique. (2022). *Pacote de Medidas de Aceleração Económica*. Ministry of Economy and Finance. Retrieved January 16, 2024, from https://www.mef.gov.mz/index.php/publicacoes/estrategias/1642-pacote-de-medidas-de-aceleracao-economica/file

Hupe, P. L., & Hill, M. J. (2016). 'And the rest is implementation.' Comparing approaches to what happens in policy processes beyond great expectations. *Public Policy and Administration*, *31*(2), 103–121. https://doi.org/10.1177/0952076715598828

INE. (2019). *Resultados definitivos. Censo de 2017. IV Recenceamento geral da população e habitação*, Instituto Nacional de Estatística.

INE. (2022). *IV General census of population and housing 2017: Socio-demographic indicators*.

Knoepfel, P., Larrue, C., Varone, F., & Hill, M. J. (2007). *Public Policy Analysis*. Policy Press

MMRE (Ministry of Mineral Resources and Energy). (2018). *Integrated master plan 2018–2043. Mozambique power system development*. Retrieved January 16, 2024, from https://www.edm.co.mz/en/document/reports/integrated-master-plan-2018-2043

MIREME. (2021). *Estatuto orgânico do ministério dos recursos minerais e energia*. Government of Mozambique. Retrieved January 16, 2024, from https://mireme.gov.mz/wp-content/uploads/2021/12/Atribuicoes-do-MIREME.pdf

Ole, N. C. (2020). The Nigerian electricity regulatory framework: Hotspots and challenges for off-grid renewable electricity development. *Journal of Energy & Natural Resources Law*, *38*(4), 367–390.

Shenga, C. & Howe, L. (2022). Oportunidades para absorver Projectos comunitários energéticos. CESET Briefing. Retrieved January 16, 2024, from https://cesetproject.com/news/new-policy-brief-mozambiques-grid-energy-regulation

SNV. (2022). *Brief status report on the new regulatory framework for access to energy in off-grid areas*. Prepared by SNV-BRILHO for ARENE in September 2022.

UNIDO. (2022). *Towards Sustainable energy for all Mozambique: Project overview, Results and lessons learned*. Presentation by UNIDO representative on November 24 2022 at the BRILHO Conference 'Off-grid energy regulatory framework: A regional perspective'.

Wirth, S. (2014). Communities matter: Institutional preconditions for community renewable energy. *Energy Policy*, *70*, 236–246. https://doi.org/10.1016/j.enpol.2014.03.021

World Bank Group. (2024). *Country climate and development report*. Retrieved January 16, 2024, from https://documents1.worldbank.org/curated/en/099113023154021937/pdf/P1771520fcab7a0930aca102c9d7107d4cc.pdf

CHAPTER 12

Conclusion

Vanesa Castán Broto

Abstract This chapter reviews the lessons from previous chapters, aiming to develop a synthetic perspective on the contribution of community energy to accelerating the transition to sustainable energy and to incorporate justice considerations into such transition. The chapter argues that community energy is a means to build strong foundations for a transition to sustainable energy, challenging the epistemic injustices embedded in current energy systems. It also argues for engaging in the practice of commoning energy as a means to address and generate action to respond to the energy justice dilemmas raised by the transition.

Keywords Community energy · Resilience · Intersectionality · Makeshift energy systems

V. Castán Broto (✉)
University of Sheffield, Sheffield, UK
e-mail: v.castanbroto@sheffield.ac.uk

V. Castán Broto (ed.), *Community Energy and Sustainable Energy Transitions*, https://doi.org/10.1007/978-3-031-57938-7_12

12.1 Introduction

This book emerged from a desire to bring coherence to the collective project of CESET, finding common threads across disciplines and perspectives. Like CESET, this book is inspired by a normative perspective on community energy: a perspective that sees community energy as a positive force within the current landscape of energy transitions. This normative commitment to community energy relates to a long tradition of environmental thought that connects goodness, peace, and sustainability to 'smallness' because small-scale interventions are less likely to cause harm to the environment (Schumacher, 1973). The importance of place-based action and the generation of collective solidarities have engendered debate on appropriate forms of socio-economic organisation, from eco-localism (Curtis, 2003) to community economies (Gibson-Graham & Roelvink, 2016). However, these attempts to redefine local economies in ecological terms struggle to engage with the promise of technology, wary of the emphasis on technological fixes that tend to pervade proposals for sustainable futures (Kerschner et al., 2018). Schumacher's 50-year-old peace-inspired vision emphasised the making of technology as a factor influencing how it mediated or prevented the unbridled exploitation of natural resources as long as technology is accessible and leaves room for human creativity (Schumacher, 1973). Community energy engages with this reflexive attention to the potential of new renewable technologies to revive places, something that counters the increasing deployment of renewables as providing technofixes for the increasingly present ecological crisis (Rudolph, 2023). But the context of community energy is uneasy as many fear its promises are mere illusions. Community energy could be likened to a hologram, a seductive three-dimensional image incorporating the multiple dimensions of energy justice, which is, in fact, nothing else than an optical illusion, beyond which lies a flat technofix. This book argues that community energy is not an illusion or a hologram; it is indeed a means to rethink the transition to sustainable energy.

However, the most destructive criticism of community energy is not its suitability to deliver sustainable energy but, rather, its futility. Community energy appears to provide nice-to-look anecdotal projects that, even if they succeed, sooner or later become overtaken by the unstoppable advance of large-scale renewables. The efforts towards community energy appear as worthless, vanity gestures to an alternative that does not really exist in the face of more efficient, better-run networks capable of reaching economies of scale. Community energy is then displaced to marginal spaces, linked to the remote, where no alternative is available. There is little recognition of the complex role that small-scale decentralised projects and cooperative forms of organisation have played in the history of electricity networks (Lehtonen & Nye, 2009). Such a sense of futility is expressed by different kinds of people, from electricity professionals to policymakers to people who could benefit from the services provided by community energy. There is a resistance to rethinking the electricity network beyond a dichotomy between centralised and decentralised systems (e.g., Bauknecht et al., 2020), only exacerbated by concerns about the impact of decentralised systems on the reliability of energy networks (Veldhuis et al., 2018). But there is also a sense of preoccupation about being disconnected, about not being part of broader projects of progress and nation-building enshrined in electrification projects. What is never spoken about enough is the failure of the modern infrastructural ideal of reliable, universal provision (Graham & Marvin, 2002), its incompleteness, its dependency on government subsidies, and the extent to which it became an instrument of control and inequity reproduction (Coutard, 2008). The globalisation of networks has not accelerated their splintering but rather highlighted their already existing precarity. Even in countries where 100% of the population has access to electricity, such as the UK, multiple factors such as the quality of the built environment, social relations, or affordability generate energy vulnerabilities manifest in poorer people's lived experiences (Middlemiss & Gillard, 2015). In engaging with the notion of infrastructure incompleteness in Nairobi, Guma (2020; p. 728) argues:

> I make the case for incompleteness as a notion that opens up a frame for analyzing a kind of urban infrastructure that, while diverging from so-called norms and ideals, cannot be described as failed or broken but as something else entirely.

This notion of emergent infrastructures in-the-making speaks to uncharted infrastructure futures requiring political alternatives. Community energy speaks to new conceptions of energy networks that, while recognising their incompleteness, emphasise the notion of autonomy. A body of research on energy sovereignty increasingly focuses on enabling the autonomy of communities to participate in decisions about their energy service (e.g., Castán Broto, 2017; Schelly et al., 2020). Community energy makes it explicit how those decisions can be advanced through a direct engagement with the technologies that provide electricity. Back to Schumacher's enchantment with technology, the technology works in small-scale projects when everybody has access to it and community creativity is prioritised. Community energy is the means to do just that.

The contributors to this book have dedicated part of their careers to understanding how to deliver community energy and in which ways it benefits those communities that engage with it. In addition, in this book, they were asked to interrogate whether those contributions extended beyond specific projects to influence a systemic change of energy systems at the regional or national levels. This conclusion aims to elaborate a synthesis of lessons learned, both in the acceleration of transitions and the manner of their delivery because it matters that transitions to sustainable energy are also fair and just.

The combined reading of the previous chapters throws two lessons for a normative perspective on community energy. The first lesson is that for community energy to advance transitions to sustainability, it requires patience: it requires being able to wait, being able to persist despite breakdowns and annoyances, and overcoming difficulties in making the project possible. The idea of patience has already emerged in previous chapters, particularly in relation to the need to find patient capital to finance community energy projects in countries like Malawi (Chapter 5). At the same time, patience is required not only from the investments and grants that support community energy but also from infrastructure managers (in making projects work) and communities (in maintaining hope over time as projects consolidate and expand). Too often, community energy projects are evaluated in terms of success whether they continue over time without recognising that failure alone is not an indicator of the extent to which a given project has had a lasting impact on the energy landscape in which it takes place, helping to consolidate ideas, launch other projects, create new expectations among communities.

The second lesson is that the know-how of community energy projects emerges from practice as practitioners have engaged with the possibilities to achieve different degrees of energy autonomy in their own communities. Community energy projects are themselves diverse, but in every case, they require assembling an array of endogenous and exogenous resources, requiring the negotiation of multiple interests and perspectives. This again calls for rethinking what a thriving community energy project looks like because community energy requires various moments of 'success,' including drawing resources for the project, enrolling the community, building the project, making sure the project is maintained over time, and finding ways to adapt the project to changing energy policies and changing energy demands. It is almost impossible for community energy projects to be successful every time and all the time. And yet, community energy projects continue to emerge out of utopian dreams of a more sustainable, resilient, and inclusive energy system. Community energy is often nothing more than an example of makeshift infrastructures: an example that helps reimagine sustainable energy futures in practice.

12.2 In Which Ways Does Community Energy Contribute to Accelerating the Transition to Sustainability?

Following climate activists' metaphors, Chapter 1 proposed to think of the transition to sustainable energy (and to sustainable futures more generally) as a massive collective project, something akin to the construction of a cathedral. That is a transition: a long term, complex project, whose completion over centuries depends on shrewd planning but also on the interactions of multiple actors, overcoming multiple difficulties within the project and beyond the project. Take for example the case of the famous basilica La Sagrada Familia in Barcelona: at its core, the basilica embodies the vision of just one person, Antoni Gaudí, but the building would not exist without the monumental collective effort sustained through private donations, public support, and professional commitments that since 1882 have sustained its construction, which will likely not be completed until 2040. Like a transition, it is a dream materialised in space. Rather than reconfiguring an existing regime the perspective that dominates current discussions of transitions to sustainability (Markard et al., 2020)—the metaphor of the cathedral invokes a different perspective away from disruption and in favour of engaging with the provisionally, incompleteness and malleability of current infrastructure systems.

At the same time, it is instructive to think of the ways in which a cathedral is not an apt metaphor for a transition, specially a just one. First, there is of course the use of an architectural metaphor from the West, which may impose certain forms of coloniality in the way transitions are approached. For example, using the mosque as a metaphor for transitions instead would emphasise horizontal, rather than vertical, expansions in the conception of a transition, perhaps turning to the flows of people through the building rather than the imposing vision of the cathedral. Which leads us to the second limitation to deploy this metaphor: the reliance on grand visions of futures, sometimes produced by a relatively small number of individuals. The transition is a collective effort, not only in the summing up of efforts but also in the combination of future visions. Here is where the exploration of a metaphor helps us to understand the focus of current efforts and what matters in activating action for a transition.

If community energy is part of the foundation of a transition, this is not because that transition is predicated on the technology advanced through community energy but because community energy helps in creating a solid ground over which different building blocks of the transition can be laid. Such solid ground consists of three crucial contributions of community energy: linking energy provision directly not to economic profits but to sustaining thriving communities, facilitating the democratisation of energy systems through the active participation of communities in their governance, and challenging the epistemic injustices that prevent access to technology. Most chapters in the book touch upon these three themes.

The book departs from a recognition of the multiple benefits of community energy to the communities who participate in such projects (Fig. 12.1). The overview of such benefits provided in Chapter 1 resonates with the practical experiences, particularly in Chapters 9 and 10 in which the accounts of projects in Malawi demonstrate how the active involvement of communities enables projects that become locally cherished and celebrated.

Chapter 2 elaborates these contributions from the perspective of building community resilience. The focus on resilience is transformative because it emphasises an entirely different model of thriving communities. The increased deployment of resilience discourse as an strategy to deliver forms of neoliberal planning that neither recognise local needs nor question the imposition of external, investment-led solutions to problems that do not exist has raised concerned critiques of a shift of emphasis from sustainability to resilience (Kaika, 2017). There is a concern that strategies advanced under the banner of resilience, from large concrete infrastructures for protection to smart city projects, not only fail to build thriving communities but also harm their long-term sustainability leading to forms of maladaptation (Eriksen et al., 2021).

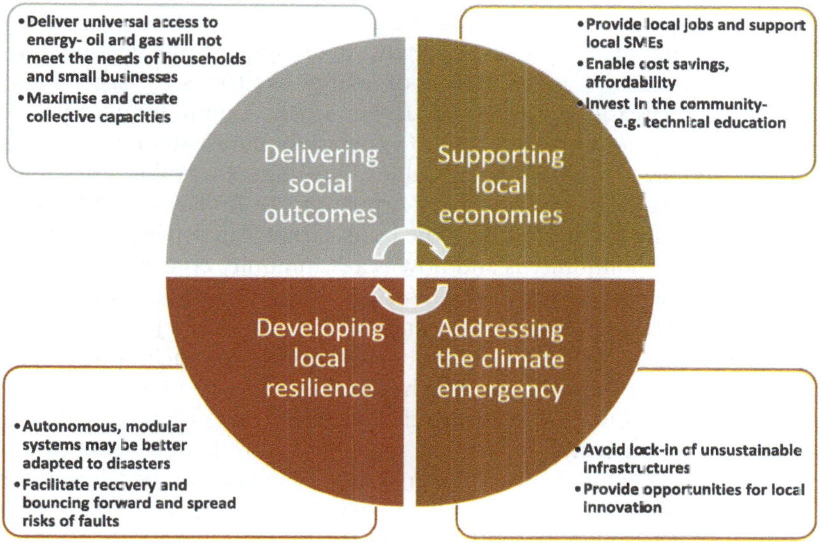

Fig. 12.1 How community energy benefits the communities that participate in those projects (*Source* own elaboration)

The approach to community resilience advanced through community energy projects shifts this thinking on its head by examining resilience as a means to transform the political systems that impact the wellbeing of communities (Pelling, 2010). This advances the perhaps subversive notion that standing your ground is for communities an instance of radical politics (Bahadur & Tanner, 2014). Chapter 2 explains that such political community resilience also depends on direct engagement with the different stages of technology implementation and development. The implicit question in community energy projects is the location of the potential for radical political change: is it in universalising assumptions of collective wellbeing that inspired infrastructural nation-building projects or on the promise for autonomy embedded in modular, flexible, and interchangeable technologies that make community energy possible?

At the same time, community energy entails an opening for the democratisation of energy systems, another element of the foundations of a transition to sustainable energy. Chapter 4 engages directly with this challenge reflecting on the challenges posed by renewable energy, and the realisation that renewable technologies do not 'automatically produce democracy and justice.' Unfortunately, renewable energy technologies enable new means of appropriation of energy territories by large energy companies, increasingly contested across space. Community energy is not exempt of contestation. Micro-generation infrastructures generate tensions and are fraught with difficulties, creating new political dynamics and reconfigure existing discourses (Armstrong & Bulkeley, 2014). That is the reason why the argument of the book is articulated around the question of what a community is and how is a community constituted around community energy (Chapter 3). This is not an easy question for the interpretations of community are many and their enactment in practice entails a politics of place: building solidarities and processes of inclusion requires assembling common purposes which necessarily create a parallel process of boundary making and implicit exclusion. However, rather than striving for perfect communities of homogeneously happy people, community energy engages with complex heterogeneous groupings which contain multiple interests. Conflict is not an extraneous negative force but a constitutive element of the formation of those communities. By building solidarities such communities are able to assemble resources (Chapter 5), to counter regulations that do not address their needs (Chapters 6, 7, and 11), and create a sense of pride that could be thought of as a means for empowerment (Chapter 10).

This goes hand in hand with communities' access to technology, which requires a good fit between communities' skills and the technical requirements of community energy. Without doubt, one of the main contributions of community energy to the foundations of transitions is challenging the epistemic injustices that prevent the development of community energy. Development organisations such as Practical Action, one of the charities within the Schumacher's circle, have proposed to have a debate on technology justice, that is, on the uneven distribution of knowledge resources which limits people's possibilities to access innovations and development opportunities (e.g., Milner, 2017). However, more recent debates on postcolonial knowledge have emphasised how imperial notions of objective knowledge, scientific rationality, and social norms have created forms of coloniality that remain relevant to understand the production of knowledge today (Tamale, 2020). This thinking resonates with an established criticism within the feminist literature about the situatedness of knowledge (raised in Chapter 4). Rather than having the right knowledge to pass to communities, the question is how existing knowledge is structured by certain imperial visions that contribute to reproduce inequalities.

There are epistemic injustices the injustices relate to the production of knowledge that relate not only to the distribution of knowledge resources across society but also with the deficits of credibility that people face because they belong to certain groups within society (e.g., marginalised, poor, colonised communities) (Fricker, 2007). Co-production projects, such as community energy projects, are powerful means to challenge those epistemic injustices (Castán Broto et al., 2022). Figure 12.2 explores how community energy projects have the potential to challenge different epistemic injustices, from those that relate to the distribution of knowledge resources to those that relate to the recognition and legitimisation of different actors to hold such knowledge. In this book we have aimed to show additional examples of those contributions, showing particularly how the experience of community projects disrupts ideas about the dominance of private actors in the energy system (e.g., Chapters 7 and 11).

However, from the outset, we dedicated specific attention to the process of building legitimacy around energy knowledge. The incursion

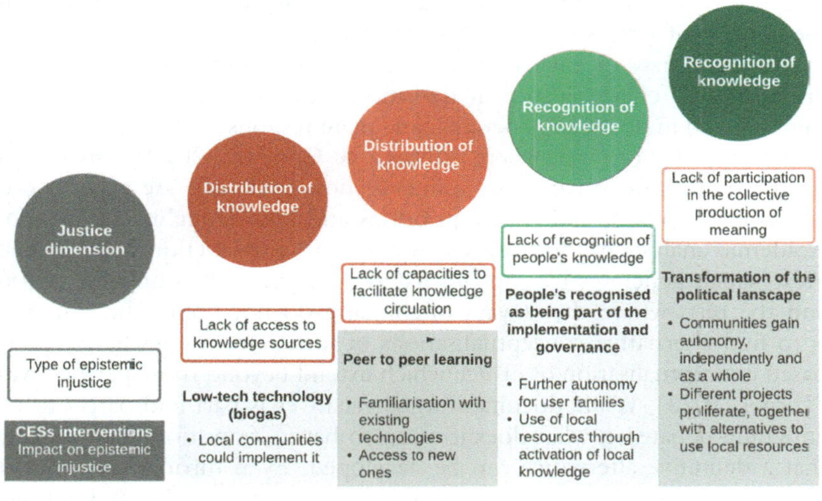

Fig. 12.2 Examples of how community energy projects address different dimensions of epistemic injustice (*Source* Castán Broto & Robin, 2023)

of this book in this territory is merely exploratory, and for now, limited to the case of Ethiopia. However, Chapter 8 shows what happens when we examine simultaneously what skills are needed for a transition to sustainable energy and what skills are offered within current higher education systems. The chapter throws light on the gender gap and the reproduction of technocratic values in the transition to sustainable energy in Ethiopia, observations that resonate with the incipient results from Malawi and Mozambique. This book was initially thought to address two audiences: community energy practitioners and policy makers. However, its main lesson is for academics: we owe it to our society to deliver a better education to build the foundations of a transition to sustainable energy.

12.3 In What Ways Does Community Energy Advance Justice in Transitions?

Achieving a transition, though, is different from doing that in a just manner. Clearly some of the foundations proposed above are indeed means to advance forms of redistribution and wellbeing, facilitate the recognition of communities' interests and perspectives, and improve the participation of all people in the transition. However, is that enough evidence of a just transition? Is that the kind of justice is demanded in the context of a massive technological change?

The discourses of environmental justice that inform the more specific discussions of energy justice in transition reproduce imperial models of thinking and in doing so generate inadequate responses to the possibilities of a just transition everywhere (Álvarez & Coolsaet, 2020). Discourses of just transitions in particular influence how trade-offs are experienced and conceptualised, and whose priorities and knowledge are prioritised. Academic analysis has focused on establishing a distinction between the West which generates ideas of justice as a structure discourse for action and the rest, where those ways of thinking do not hold. The response is to find alternative conceptualisations of justice from elsewhere, place-based conceptualisation of justice which extend beyond those perspectives (Tornel, 2023). While documenting alternatives is part and parcel of an enriching debate on what does it mean to have a transition, it is unlikely that a definitive alternative can be developed. Even through the lens of a variety of radical theory, energy justice does not become dramatically redefined only nuanced (see a recent debate on this topic: Dunlap & Tornel, 2023; Sovacool et al., 2023).

Perhaps the separation between two worlds, one Global South and one Global North, as shaped by a boundary across which ideas travel is itself at the root of the problem. Not only is based on a colonial misnomer that does not correspond to any identifiable geography, but also, it tends to obscure how coloniality emerge from an imperial project of colonisation that affected the whole world. For example, drawing a Global North that includes North America automatically erodes the history of First Nations in those territories. For indigenous peoples, justice emerges from the recognition of responsibilities as they are distributed across social, political, ecological, and material relations (Whyte, 2013). Finding common challenges across geographical locations and develop vocabularies to reflect the particularities of specific contexts (as seen by the communities leading energy projects) is a better strategy that confining certain vocabularies to predetermined territories. This conclusion resonates with the thesis of this book, about the need for research and learning emerging from practical experiences in a variety of countries, but specially those that have received less attention in the energy transitions literature (such as Ethiopia, Malawi, and Mozambique).

In the context of CESET, community energy emerges from a promise of engagement within communities that wish to be recognised in their complexity. Projects like Chipopoma, in North Malawi (Chapter 10) battled scepticism before the project could be established. Many local residents were not supportive of the project, but once people saw that their neighbours got electricity, they became interested in being part of it. So, while at the beginning of the project the question was whether the community was accepting of the project, as the project consolidated the question became how the project can reach everyone. That is the reason why growing and extending the network becomes so important for more mature community energy projects, as it has happened in the case of Bondo and the Mulanje National Park (Chapter 10). Justice emerges in relations that extend beyond questions of responsibility, not only through what the project provides today, but also through what it can offer in the future.

For that reason, justice becomes, most of all, a question of possibility and opening. Addressing epistemic injustices is important, in the first place, because it creates possibilities that otherwise remained closed. Possibility, however, is not only a question of knowledge. It is also a question of politics, frameworks, technologies, and resources as the chapters of the book show. In this context the book's aim is to inspire initiatives to make community energy, because it is only through the making of a collective, shared energy project that the possibilities of community energy become available, even if the project raises new justice dilemmas. The shadow of global capitalism lures in a landscape of multiple forms of precarity, but in specific locations, precarity is engaged in practical ways, engaging with whatever possibilities are provided in existing ecosystems (Tsing, 2015). Those are the possibilities over which the foundations of a transition can be built. For the communities that actively assemble their own energy projects, the question is not one of refining the definition of justice, but of constructing the means of life in whatever ruins of imperial pasts we find ourselves.

References

Álvarez, L., & Coolsaet, B. (2020). Decolonizing environmental justice studies: A Latin American perspective. *Capitalism Nature Socialism, 31*(2), 50–69.

Armstrong, A., & Bulkeley, H. (2014). Micro-hydro politics: Producing and contesting community energy in the North of England. *Geoforum, 56*, 66–76.

Bahadur, A., & Tanner, T. (2014). Transformational resilience thinking: Putting people, power and politics at the heart of urban climate resilience. *Environment and Urbanization, 26*(1), 200–214.

Bauknecht, D., Funcke, S., & Vogel, M. (2020). Is small beautiful? A framework for assessing decentralised electricity systems. *Renewable and Sustainable Energy Reviews, 118*, 109543.

Castán Broto, V. (2017). Energy sovereignty and development planning: The case of Maputo, Mozambique. *International Development Planning Review, 39*(3), 229–248.

Castán Broto, V., Ortiz, C., Lipietz, B., Osuteye, E., Johnson, C., Kombe, W., Mtwangi-Limbumba, T., Macías, J. C., Desmaison, B., & Hadny, A. (2022). Co-production outcomes for urban equality: Learning from different trajectories of citizens' involvement in urban change. *Current Research in Environmental Sustainability, 4*, 100179.

Castán Broto, V. & Robin E. (2023). Energy justice as a new communal project? Community energy systems and the energy access gap. In *Handbook on Energy Justice* (pp. 253–268). Edward Elgar Publishing

Coutard, O. (2008). Placing splintering urbanism: Introduction. *Geoforum, 39*(6), 1815–1820.

Curtis, F. (2003). Eco-localism and sustainability. *Ecological Economics, 46*(1), 83–102.

Dunlap, A., & Tornel, C. (2023). Pluralizing energy justice? Towards cultivating an unruly, autonomous and insurrectionary research agenda. *Energy Research & Social Science, 103*, 103217.

Eriksen, S., Schipper, E. L. F., Scoville-Simonds, M., Vincent, K., Adam, H. N., Brooks, N., Harding, B., Lenaerts, L., Liverman, D., Mills-Novoa, M., & Mosberg, M. (2021). Adaptation interventions and their effect on vulnerability in developing countries: Help, hindrance or irrelevance? *World Development, 141*, 105383.

Fricker, M. (2007). *Epistemic injustice: Power and the ethics of knowing*. Oxford University Press.

Gibson-Graham, J. & Roelvink, G. (2016). Social innovation for community economies. In *Social innovation and territorial development* (pp. 25–38). Routledge

Graham, S., & Marvin, S. (2002). *Splintering urbanism: Networked infrastructures, technological mobilities and the urban condition*. Routledge.

Guma, P. K. (2020). Incompleteness of urban infrastructures in transition: Scenarios from the mobile age in Nairobi. *Social Studies of Science, 50*(5), 728–750.

Kaika, M. (2017). 'Don't call me resilient again!': The New Urban Agenda as immunology... or... what happens when communities refuse to be vaccinated with 'smart cities' and indicators. *Environment and Urbanization, 29*(1), 89–102.

Kerschner, C., Wächter, P., Nierling, L., & Ehlers, M.-H. (2018). Degrowth and Technology: Towards feasible, viable, appropriate and convivial imaginaries. *Journal of Cleaner Production, 197*, 1619–1636.

Lehtonen, M., & Nye, S. (2009). History of electricity network control and distributed generation in the UK and Western Denmark. *Energy Policy, 37*(6), 2338–2345.

Markard, J., Geels, F. W., & Raven, R. (2020). Challenges in the acceleration of sustainability transitions. *Environmental Research Letters, 15*(8), 081001.

Middlemiss, L., & Gillard, R. (2015). Fuel poverty from the bottom-up: Characterising household energy vulnerability through the lived experience of the fuel poor. *Energy Research & Social Science, 6*, 146–154.

Milner, T. (2017). From intermediate technology to technology justice. In *The Social Realities of Knowledge for Development*. Practical Action 108. IDS.

Pelling, M. (2010). *Adaptation to climate change: From resilience to transformation*. Routledge.

Rudolph, D. (2023). The question of 'sustainable' technology: From socio-ecological fixes to transformations. *Human Geography, 16*(1), 81–86.

Schelly, C., Bessette, D., & D., Brosemer, K., Gagnon, V., Arola, K. L., Fiss, A., Pearce, J. M., & Halvorsen, K. E. (2020). Energy policy for energy sovereignty: Can policy tools enhance energy sovereignty? *Solar Energy, 205*, 109–112.

Schumacher, E. F. (1973). *Small is beautiful: Economics as if people mattered*. Blond & Briggs.

Sovacool, B. K., Bell, S. E., Daggett, C., Labuski, C., Lennon, M., Naylor, L., Klinger, J., Leonard, K., & Firestone, J. (2023). Pluralizing energy justice: Incorporating feminist, anti-racist, Indigenous, and postcolonial perspectives. *Energy Research & Social Science, 97*, 102996.

Tamale, S. (2020). *Decolonization and Afro-feminism*. Daraja Press.

Tornel, C. (2023). Decolonizing energy justice from the ground up: Political ecology, ontology, and energy landscapes. *Progress in Human Geography, 47*(1), 43–65.

Tsing, A. L. (2015). *The mushroom at the end of the world: On the possibility of life in capitalist ruins*. Princeton University Press

Veldhuis, A. J., Leach, M., & Yang, A. (2018). The impact of increased decentralised generation on the reliability of an existing electricity network. *Applied Energy, 215*, 479–502.

Whyte, K. P. (2013). Justice forward: Tribes, climate adaptation and responsibility. In *climate change and indigenous peoples in the United States: Impacts, experiences and actions* (pp. 9–22). Springer International Publishing

Index

V. Castán Broto (ed.), *Community Energy and Sustainable Energy Transitions*, https://doi.org/10.1007/978-3-031-57938-7

The manufacturer's authorised representative in the EU is Springer Nature Customer Service Centre GmbH, Europaplatz 3, 69115 Heidelberg, Germany. If you have any concerns regarding our products, please contact ProductSafety@springernature.com

Printed and bound by CPI Group (UK) Ltd, Croydon, CR0 4YY
30/09/2026
02225573-0001